AF516172

Encyclopedia of Nanotechnology

Encyclopedia of Nanotechnology

Dr. S.K. Singh

AMIGA PRESS. INC
Delhi -110051

AMIGA PRESS. INC
H.O :- 28-A, Indra Park, Chander Nagar
Delhi :- 110051 (India)
Tel :- 011-22504028, +91-98736-57000
Email :- amigapress@gmail.com
Website: www.amigapress.co.in

This edition is published by Amiga Press.Inc

ISBN : 978-93-84533-06-9

Copyright© Reserved
First Published 2018

Publisher's note:
This publication may not reproduced, stored in a retrieval system or transmitted in any form or by any means, photocopying , mechanical, electronic, recording or otherwise, without the prior permission from publication house.

Printed and bound in India by Thomson Press Ltd., New Delhi

Preface

Nanotechnology is the science and creation of materials at a molecular level. Nanotechnology applications are used in medicine, chemistry, the environment, energy companies, and communication. Whether it is through nanotechnology education or nanotechnology application, this science continues to evolve and find more uses in everyday life. The use of nanotechnologies in medicine has provided many medical breakthroughs. Some of the medical nanotechnology applications are for diagnostics. These are used to assist in identifying DNA and helping detect genetic sequences. Other uses of nanotechnology in medicine are in drug delivery and tissue engineering.

Nanomanufacturing is used to describe either the production of nano scaled materials, which can be powders or fluids, or describe the manufacturing of parts "button up" from nanoscaled materials or "top down" in smallest steps for high precison by several technologis like laser ablation, etching and others. The term is widely used, e.g. by the European Technology Platform MINAM and the U.S. National Nanotechnology Initiative (NNI). NNI refer to a sub-effort of nanotechnology--one of its five "priority areas."

Drug delivery can help in cancer treatments by administering drugs via implant versus the traditional injections. Tissue engineering is a new nanotechnology and involves repairing or artificially reproducing damaged tissue. Using molecular nanotechnology in chemistry and the environment has already resulted in positive benefits. Chemical catalysis is a process of reducing pollutants using a reagent or catalyst. This can be especially helpful if used in vehicle fuel cells or catalytic converters to reduce car emissions. Filtration is another one of the nanotechnology applications that can help the environment. Extremely small holes in the filtration, known as nanofiltration, remove ions for wastewater treatment or air purification. Slightly larger holes are used in a type of filtration called ultra filtration and its current application is for renal dialysis.

A nanostructure is an object of intermediate size between molecular and microscopic (micrometer-sized) structures. In describing nanostructures it is necessary to differentiate between the number of dimensions on the nanoscale. Nanotextured surfaces have one dimension on the nanoscale, i.e., only the thickness of the surface of an object is between 0.1 and 100 nm. Nanotubes have two dimensions on the nanoscale, i.e., the diameter of the tube is between 0.1 and 100 nm; its length could be much greater. Finally, spherical nanoparticles have three dimensions on the nanoscale, i.e., the particle is between 0.1 and 100 nm in each spatial dimension.

Nanotechnology can be even broader than that. A rough rule of thumb is: if it's too small to see with an ordinary light microscope, it's likely to be considered nanotechnology. Without using special physics tricks, light can't be used to see anything smaller than half a wavelength of light, which is a few hundred nanometers. Because some optics technology uses structures smaller than light (such as photonic crystals) to manipulate light, you will sometimes see optics researchers describe their work as nanotechnology. However, because these structures tend to be larger than the official 100-nm cutoff, many nanotechnologists will reject this usage. Another point of contention is how unique the "unique phenomena enabl[ing] novel applications" have to be. For example, some nanotechnology simply uses ordinary materials like clay, in smaller chunks, in fairly ordinary ways. They can get new material properties; they are using nanoscale materials; they are studying them with new techniques; but is it really nanotechnology, or is it just materials science? It might as well be called nanotech, seems to be the consensus. It's providing early successes for the field and it's putting "nano" into consumers' hands in a beneficial, non-threatening way.

The book will serve as an ultimate resource book for everyone who would like to know the wide range of prospects in the emerging applications of Nanotechnology.

—Editor

Contents

1

Introduction

Nanotechnology is the study and use of structures between 1 nanometer and 100 nanometers in size. To give you an idea of how small that is, it would take eight hundred 100 nanometer particles side by side to match the width of a human hair.

Scientists have been studying and working with nanoparticles for centuries, but the effectiveness of their work has been hampered by their inability to see the structure of nanoparticles. In recent decades the development of microscopes capable of displaying particles as small as atoms has allowed scientists to see what they are working with.

The following illustration titled "The Scale of Things" was created by the U. S. Department of Energy. The chart starts with objects that can be seen by the unaided eye, such as an ant, at the top of the chart, and progresses to objects about a nanometer or less in size, such as the ATP molecule used in humans to store energy from food. Viewing these nano-sized objects requires powerful microscopes.

The ability to see nano-sized materials can make a huge difference to researchers. Consider the challenge of drug delivery. Today, most harmful side effects of treatments such as chemotherapy are a result of drug delivery methods that don't pinpoint their intended target cells accurately. Researchers at Harvard and MIT have been able to fill nanoparticles with special RNA strands sticking out from their surface with a chemotherapy drug. These RNA strands are attracted to cancer cells. When the nanoparticle encounters a cancer cell it adheres to it and releases the drug into the cancer cell. This directed method of drug delivery has great potential for treating cancer patients, producing less side harmful affects than those produced by conventional chemotherapy. Imagine researchers without the use of microscopes capable of resolving the structure of nano-particles and cells doing this work; it would be like assembling a complex model blindfolded. The properties of familiar materials, such as silver or iron, are being changed by manufacturers who are adding nano-sized components to conventional materials to improve performance. For example, some clothing manufacturers are making water and stain repellent clothing using nano-sized whiskers in the fabric that cause water to bead up on the surface. The properties of many conventional materials change when created with nano-sized particles (nanoparticles). This is generally because nanoparticles have a greater surface area per weight than larger particles; they are therefore more reactive to some other molecules. For example studies have show that nanoparticles of iron can be effective in the cleanup of chemicals in groundwater because they react more efficiently to those chemicals.

Nano-sized particles of carbon, (for example nanotubes and bucky balls) are extremely strong. Nanotubes and bucky balls are composed of only carbon and their strength comes from special characteristics of the bonds between carbon atoms. One proposed application that illustrates the strength of nanosized particles of carbon is the manufacture of t-shirt weight bullet proof vests made out of carbon nanotubes.

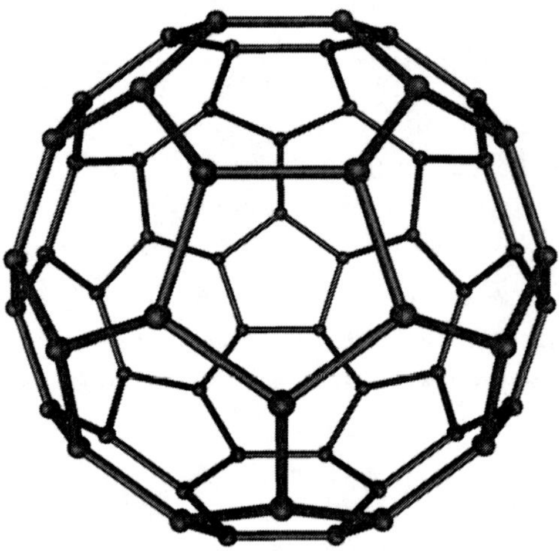

Fiure: ***Buckminsterfullerene C_{60}, also known as the buckyball, is a representative member of the carbon structures known as fullerenes. Members of the fullerene family are a major subject of research falling under the nanotechnology umbrella.***

The ability to create gears, mirrors, sensor elements, as well as electronic circuitry in silicon surfaces allows the manufacture of miniature sensors such as those used to activate the air bags in your car. This technique is called MEMS (Micro-ElectroMechanical Systems) The MEMS technique results in close integration of the mechanical mechanism with the necessary electronic circuit on a single silicon chip, similar to the method used to produce computer chips. Using MEMS to produce a device reduces both the cost and size of the product, compared to similar devices made with conventional methods.

A vision for the future is the construction of molecular machines that would be constructed atom by atom to build certain desirable properties into a device itself, rather than etching away solid silicon to form the desired mechanism, as is the case in the MEMS process.

TYPES OF NANOTECHNOLOGY

Nanotechnology is the science and creation of materials at a molecular level. Nanotechnology applications are used in medicine, chemistry, the environment, energy companies, and communication. Whether it is through nanotechnology education or nanotechnology application, this science continues to evolve and find more uses in everyday life. The use of nanotechnologies in medicine has provided many medical breakthroughs. Some of the medical nanotechnology applications are for diagnostics. These are used to assist in identifying DNA and helping detect genetic sequences.

Other uses of nanotechnology in medicine are in drug delivery and tissue engineering. Drug delivery can help in cancer treatments by administering drugs via implant versus the traditional injections. Tissue engineering is a new nanotechnology and involves repairing or artificially reproducing damaged tissue. Using molecular nanotechnology in chemistry and the environment has already resulted in positive benefits. Chemical catalysis is a process of reducing pollutants using a reagent or catalyst. This can be especially helpful if used in vehicle fuel cells or catalytic converters to reduce car emissions. Filtration is another one of the nanotechnology applications that can help the environment. Extremely small holes in the filtration, known as nanofiltration, remove ions for wastewater treatment or air purification. Slightly larger holes are used in a type of filtration called ultra filtration and its current application is for renal dialysis.

Now that nanotechnology has been in the public eye for twenty years, and well-funded for half a decade, it's worth a quick look at just what it is—and how it got that way. When the word "nanotechnology" was introduced to the public by Eric Drexler's 1986 book *Engines of Creation,* it meant something very specific: small precise machines built out of molecules, which could build more molecular machines and products—large, high-performance products. This goal or aspect of nanotechnology now goes by several names, including molecular nanotechnology, molecular manufacturing, and productive nanosystems. The reason for this renaming is that "nanotechnology" has become a broad and inclusive term, but it's still important to distinguish molecular manufacturing from all the other types. I'll talk about molecular manufacturing, and why it is unique and important, after surveying some of the other types of nanotechnology.

With the funding of the U.S. National Nanotechnology Initiative (NNI), there has been a strong financial incentive to define nanotechnology so that one's own research counts—but not so broadly that everyone's research counts. There has been a less focused, but still real, incentive to define the goals of nanotechnology aggressively, to justify major funding, but not too aggressively, lest it sound scary or implausible.

With all the different research fields applying the above rules to a wide variety of research, it is not surprising that there's no single hard-edged definition of nanotechnology that everyone can agree on. Perhaps the most commonly quoted definition of nanotechnology is the one used by the NNI: "Nanotechnology is the understanding and control of matter at dimensions of roughly 1 to 100 nanometers, where unique phenomena enable novel applications." I don't know how they decided on the size scale; thinking cynically, it might have had something to do with the fact that computer chips were just about to gain features smaller than 100 nanometers, so they were guaranteed at least one early success.

Nanotechnology can be even broader than that. A rough rule of thumb is: if it's too small to see with an ordinary light microscope, it's likely to be considered nanotechnology. Without using special physics tricks, light can't be used to see anything smaller than half a wavelength of light, which is a few hundred nanometers. Because some optics technology uses structures smaller than light (such as photonic crystals) to manipulate light, you will sometimes see optics researchers describe their work as nanotechnology. However, because these structures tend to be larger than the official 100-nm cutoff, many nanotechnologists will reject this usage. Another point of contention is how unique the "unique phenomena enabl[ing] novel applications" have to be. For example, some nanotechnology simply uses ordinary materials like clay, in smaller chunks, in fairly ordinary ways. They can get new material properties; they are using nanoscale materials; they are studying them with new techniques; but is it really nanotechnology, or is it just materials science? It might as well be called nanotech, seems to be the consensus. It's providing early successes for the field and it's putting "nano" into consumers' hands in a beneficial, non-threatening way.

Another kind of nanotechnology involves building increasingly large and intricate molecules. Some of these molecules can be very useful: for example, it appears possible to combine a cancer-cell-recognizer, a toxic drug, and a component that shows up in MRI scans, into a single molecule that kills cancer cells while showing you where they were and leaving the rest of the body untouched. This is a little bit different from traditional chemistry in that the chemist isn't trying to create a new molecule with a single function, but rather to join together several different functions into one connected package.

Some new nanomaterials have genuinely new properties. For example, small mineral particles can be transparent to visible light, which makes them useful in sunscreen. Even smaller particles can glow in useful colors, forming more-stable markers for biomedical research. For related reasons, small particles can be useful additions to computer circuits, lending their quantum effects to make smaller and better transistors.

We should talk about semiconductors (computer chips), a major application of nanotechnology. Feature sizes on mainstream silicon chips are well below 100 nanometers now. This obviously is a great success for nanotechnology (as defined by the NNI). From one point of view, semiconductor makers are continuing to do what they have always done: make chips smaller and faster using silicon-based transistors. From another point of view, as sizes shrink, their task is rapidly getting harder, and they are inventing new technology every day just to keep up with expectations. There are more unusual computer-chip designs underway as well, most of which use nanotechnology of one form or another, from quantum-dot transistors to sub-wavelength optics (plasmonics) to holographic storage to buckytube-based mechanical switches. Which brings us to buckytubes. Buckytubes are remarkable molecules that were discovered not long ago. They are tiny strips of graphite, rolled up with the sides fastened together to form a seamless tube. They are very strong, very stiff, and can be quite long in proportion to their width; four-centimeter long buckytubes have been reported, which is more than ten million times the width of the tube. Some buckytubes are world-class conductors and electron emitters. They may be useful in a wide variety of applications.

And what about those quantum effects? According to the NNI, "At the nanoscale, the physical, chemical, and biological properties of materials differ in fundamental and valuable ways from the properties of individual atoms and molecules or bulk matter." Materials are of course made up of atoms, which contain electrons, and it is the interaction of electrons that gives materials most of their properties. In very small chunks of material, the electrons interact differently, which can create new material properties. Nanoparticles can be more chemically active; as mentioned above, they can fluoresce; they can even participate in weird physics such as quantum computers. But, as the above overview should make clear, a lot of "nanotechnology" does not make use of these quantum effects.

Molecular manufacturing (MM) is a fairly mundane branch of nanotech, or it would be if not for the political controversy that has swirled around it. The idea is simple: Use nanoscale machines as construction tools, joining molecular fragments into more machines. Every biological cell contains molecular machines that do exactly that. There are, however, a few reasons why molecular manufacturing has been highly controversial.

Much of the controversy stems from the fact that MM proposes to use engineered devices to build duplicate devices. Although biology can do this, intuition suggests that such self-duplication requires some special spark of complexity or something even more numinous: surely a simple engineered machine can't be so lifelike! This ultimate spark of vitalism is fading as we learn how machinelike cellular molecules actually are, and as increasingly detailed plans make it clear that hardware does not have to be very complex in order to make duplicate hardware. (Even the software doesn't have to be very complex, just intricate and well-designed. This has been known by computer scientists for many decades, but the paradigm has taken a while to shift in the wider world.)

There is another problem with self-replication: in some forms, it may be dangerous. In 1986, Eric Drexler warned that tiny engineered self-replicators could outcompete natural life, turning the biosphere into boring copies of themselves: "grey goo." This formed a cornerstone of Bill Joy's essay "Why The Future Doesn't Need Us," which hit just as the NNI was ramping up. No nanoscientist wanted to be associated with a poorly-understood technology that might destroy the world, and the easiest thing was to assert that MM was simply impossible. (Modern MM designs do not use small self-replicators; in fact, they have been obsolete since Drexler's 1992 technical book *Nanosystems*.)

A third source of controversy is that MM plans to use diamond as its major building material, not bio-based polymers like protein and DNA. (Some pathways to this capability, including the pathway favored by Drexler, go through a biopolymer stage.) Although there is a wide variety of reactions that can form diamond and graphite, living organisms do not build with these materials, so there is no existence proof that such structures can be built using point-by-point computer-controlled molecular deposition. If diamond-like structures can be built by molecular manufacturing techniques, they should have astonishingly high performance characteristics. To those who study MM, its projected high performance indicates that researchers should work toward this goal with a focused intensity not seen since the Manhattan Project. To those who have not studied MM, talk of motors a million times more powerful than today's merely seems fanciful, a reason (or an excuse) to discount the entire field.

At least as problematic as the extreme technical claims are the concerns about the extreme implications of molecular manufacturing. It is rare that a technology comes along which revolutionizes society in a decade or so, and even more rare that such things are correctly predicted in advance. It is very tempting to dismiss claims of unstable arms races, wholesale destruction of existing jobs, and widespread personal capacity for mass destruction, as improbable.

However, all the skepticism in the world won't change the laws of physics. In more than two decades (almost five, if you count from Richard Feynman's visionary speech), no one has found a reason why MM, even diamond-based MM, shouldn't work. In fact, the more work that's done, the less complex it appears. Predicting social responses to technology is even more difficult than predicting technology itself, but it seems beyond plausibility that such a powerful capability won't have at least some disruptive effects—perhaps fatally disruptive, unless we can understand the potential and find ways to bypass the worst pitfalls.

In the near future, nanotechnology in the broad sense will continue to develop dozens of interesting technologies and capabilities, leading to hundreds of improved capabilities and applications. Meanwhile, molecular manufacturing will continue to move closer, despite the (rapidly fading) opposition to the idea. Sometime in the next few years, someone will have the vision to fund a targeted study of molecular manufacturing's potential; less than a decade after that, general-purpose nanoscale manufacturing will be a reality that the world will have to deal with. Molecular manufacturing will build virtually unlimited quantities of new products as rapidly as the software can be designed—and it should be noted that most of today's physical products are far less complex than today's software. Molecular manufacturing will both enable and eclipse large areas of nanotechnology, further accelerating the achievements of the field. We are in for some interesting times.

BOTTOM-UP DESIGN

At first encounter, the idea of designing products with 100,000,000,000,000,000,000,000 atoms, each in an engineered position, and each one placed without error, may seem ridiculous. But the goal is not as implausible as it sounds. Today's personal computers do that number of transistor operations every few weeks. The operations are done without error, and each one was engineered—though not directly. There are two reasons why computers can do this: digital operations and levels of abstraction. I've talked about both of these in previous essays, but it bears repeating: at the lowest level of operations, personal computers do a mole of engineered, reliable transistor operations every few weeks, and the techniques used to accomplish this can be applied to molecular manufacturing.

Computers can be so precise and reliable because they are based on digital operations. A digital operation uses discrete values: either a 1 or a 0. A value of 0.95 will be corrected to 1, and a value of 0.05 will be corrected to 0. This correction happens naturally with every transistor operation. Transistors can do this correction because they are nonlinear: there is a large region of input where the output is very close to 1, and another large region of input where the output is very close to 0. A little bit of energy is used to overcome entropy at each step. Rather than letting inaccuracies accumulate into errors, they are fixed immediately. Thermal noise and quantum effects are corrected before they compound into errors.

Forces between atoms are nonlinear. As atoms approach each other, they feel a weak attractive force. Then, at a certain distance, the force becomes repulsive. If they are pushed together even more closely, the force becomes more strongly attractive than before; finally, it becomes sharply repulsive. Chemists and physicists know the region of weak distant attraction as "surface forces"; the closer, stronger attraction is "covalent bonds"; and the intervening zone of repulsion is responsible for the "activation energy" that is required to make reactions happen. Of course, this picture is over-simplified; covalent bonds are not the only type of bond. But for many types of atoms, especially carbon, this is a pretty good description.

Several types of errors must be considered in fabricating and using a mechanical component. A fabrication operation may fail, causing the component to be damaged during manufacture. The operations may be correct but imprecise, causing small variances in the manufactured part. During use, the part may wear, causing further variance. As we will see, the nonlinear nature of molecular bonds can be used (with good design) to virtually eliminate all three classes of error.

Nonlinear forces between atoms can be used to correct inaccuracies in fabrication operations before they turn into errors. If the atom is placed in slightly the wrong location, it will be pulled to the correct location by inter-atomic forces. The correction happens naturally. If the placement tool is inaccurate, then energy will be lost as the atom moves into place; as with transistors, entropy isn't overcome for free. But, as with transistors, reliability can be maintained over virtually unlimited numbers of operations by spending a little bit of energy at each step.

In practice, there are several different kinds of errors that must be considered when a moiety—an atom or a molecular fragment—is added to a part under construction. It may fail to transfer from the "tool" molecule to the "workpiece" molecule. This kind of error can be detected and the operation can be retried. The moiety may bond to the wrong atom on the workpiece. Or it may exert a force on the workpiece that causes other atoms, already in the workpiece, to rearrange their bonds. This is called "reconstruction," and avoiding it imposes additional requirements for precise placement of the moiety, but it is also a non-linear phenomenon: if the moiety is positioned within a certain range of the ideal location, reconstruction won't happen, at least in well-chosen structures.

Errors of dimensional tolerance, which in traditional manufacturing are caused by imprecise operations or wear during operation, need not be a factor in molecular manufactured components. If an atom is pulled slightly out of place, either during manufacture or during operation, it will be pulled back into place by its bonds. In engineering terms, there is no plastic deformation, only elastic deformation. Of course, if a strong enough force is applied, the bonds can be broken, but preventing this is a matter of engineering the product properly. It requires a lot of force to break a bond. If a component must be either perfectly fastened or broken, then it will remain perfect for a long, long time under normal usage.

Traditional mechanical engineering and manufacturing involve a lot of operations to deal with errors of dimensional tolerance—including measuring, finishing, and sensing during operation—that will not be required with molecular manufactured components. This will make molecular manufacturing systems significantly easier to automate. As long as low-level operations are reliable and repeatable, then higher-level operations built on them also will be reliable. Knowing precisely how the system works at the lowest level will allow confident engineering at higher levels. This design principle is called levels of abstraction.

A computer programmer can write an instruction such as, "Draw a black rectangle in the middle of the screen," in just a few characters of computer code. These few characters, however, may invoke thousands of low-level instructions carried out by billions of transistor operations. The programmer has implicit confidence that each transistor will work correctly. Actually, programmers don't think about transistors at all, any more than you think about each spark in your car's engine when you step on the gas. Transistors are combined into registers, which are used by CPU microcode, which is controlled by assembly language, which is machine-generated from high-level languages, which are used to write several layers of operating system functions and libraries, and this is what the programmers actually use. Because transistors are, in effect, completely reliable and predictable, each level built on top of them also is completely reliable and predictable (with the exception of design errors).

Molecular manufacturing will involve massive numbers of simple mechanosynthetic operations done under fully automated control. A nanofactory building a product would not be much different, at several important levels of function, from a computer-driven printer printing a page. The nanofactory product designer would not see each atom, any more than a graphic artist sees each ink droplet. Graphic artists usually work in abstractions such as splines, rather than individual pixels. The user does not even see each spline. The user just hits "Print" and the picture comes out of the printer with each ink droplet in its proper place. A molecular manufactured product could include a microscopic component containing a billion atoms—which could be placed with complete reliability by a single instruction written by a designer. An array of a billion identical components, each containing a billion atoms, could be specified without any additional difficulty.

Each component could reliably work for many years without a single broken bond. Thus, just as a computer programmer can write a simple program that does an almost unlimited number of reliable calculations, a product designer could write a simple specification that placed an almost unlimited number of atoms—reliably and predictably—making exactly the product that was desired. (Background radiation is beyond the scope of this essay; it will introduce failures and require redundancy at scales larger than about a micron, but this should not require much additional complexity.)

Operation of the manufactured product can be similarly planned from the bottom up. If the smallest operations happen in a predictable way at a predictable time, then higher-level operations can be built on top of the low-level functionality. This is not the only way to implement high-level functionality, of course. Biology uses statistical processes and analog feedback loops to implement its actions. Although this is more elegant and efficient in some ways, it would be difficult to design systems that worked along these lines, and it is not necessary. Digital operations can be made to happen in lockstep, and aggregates of digital operations can be treated as reliable primitives for higher levels. The more predictable a system is, the less sensing is required to make it work as desired. Nanoscale sensing often is cited as a weak point in nanomachine design, but in principle, nanomachines designed on digital principles would not need any sensing in order to work reliably. In practice, only a small amount of internal feedback would be required, which could be provided by relatively crude sensors.

It is important to realize that digital design using levels of abstraction does not imply increased complexity at higher levels. An assembly language instruction that causes a billion transistor operations may be specified completely with a paragraph of description. Its results may be very intricate—may invoke a lot of diverse activity—but there is a crucial distinction between intricacy and complexity. Similarly, a high-level language instruction that invokes a billion assembly language instructions may be understood completely at a glance. And so it goes, through as many levels as are useful to the programmer/designer. As long as the lower levels are reliable, the upper levels can be reliable, intricate (useful), and simple (easy to use). One of the most important features of molecular manufacturing is that its very lowest level—

the formation of molecules from precisely positioned building blocks—is precise and reliable due to digital operations. Every level of abstraction above the foundation of molecular fabrication can thus be equally precise and reliable. Google, the World Wide Web, and modern video games all have been engineered from molar numbers of transistor operations. In the same way, masses of diverse, highly functional products will be engineered from molecular fabrication operations.

NANOTECHNOLOGY IN SPACE

Nanotechnology may hold the key to making spaceflight more practical. Advancements in nanomaterials make lightweight solar sails and a cable for the space elevator possible. By significantly reducing the amount of rocket fuel required, these advances could lower the cost of reaching orbit and traveling in space. In addition, new materials combined with nanosensors and nanorobots could improve the performance of spaceships, spacesuits, and the equipment used to explore planets and moons, making nanotechnology an important part of the 'final frontier.'

Nanotechnology Applications under Development for Space

Researchers are looking into the following nanotechnology project in space:

- Using carbon nanotubes to make the cable needed for the space elevator, a system which could significantly reduce the cost of sending material into orbit. Nova has a nice video explaining the concepts.

A space elevator is essentially a long cable extending from our planet's surface into space with its center of mass at geostationary Earth orbit (GEO), 35,786 km in altitude. Electromagnetic vehicles traveling along the cable could serve as a mass transportation system for moving people, payloads, and power between Earth and space.

Current plans call for a base tower approximately 50 km tall--the cable would be tethered to the top. To keep the cable structure from tumbling to Earth, it would be attached to a large counterbalance mass beyond geostationary orbit, perhaps an asteroid moved into place for that purpose. "The system requires the center of mass be in geostationary orbit," said Smitherman. "The cable is basically in orbit around the Earth."

Four to six "elevator tracks" would extend up the sides of the tower and cable structure going to platforms at different levels. These tracks would allow electromagnetic vehicles to travel at speeds reaching thousands of kilometers-per-hour.

Conceptual designs place the tower construction at an equatorial site. The extreme height of the lower tower section makes it vulnerable to high winds.

An equatorial location is ideal for a tower of such enormous height because the area is practically devoid of hurricanes and tornadoes and it aligns properly with geostationary orbits (which are directly overhead).

According to Smitherman, construction is not feasible today but it could be toward the end of the 21st century. "First we'll develop the technology," said Smitherman. "In 50 years or so, we'll be there. Then, if the need is there, we'll be able to do this. That's the gist of the report."

How is nanotechnology used in space? We are currently exploring only a tiny fraction of its space capabilities as we study the development of newer, better space materials. With materials like these we may be able to find ways of launching into orbit that don't involve costly rockets (and their costly fuel.) Researchers are particularly excited about the possibilities for a space elevator.

Carbon nanotubes are the perfect choice for such an elevator's cable, since nanotechnology is able to create carbon-based material that is light in weight yet strong enough to withstand the forces it would face in space. A space elevator would make all kinds of pioneering efforts possible by dramatically reducing the cost of sending things into orbit.

Figure: ***Applications of Nanotechnology in Space Developments and Systems***

Smitherman's paper credits Arthur C. Clarke with introducing the concept to a broader audience. In his 1978 novel, Fountains of Paradise, engineers construct a space elevator on top of a mountain peak in the mythical island of Taprobane (closely based on Sri Lanka, the country where Clarke now resides). The builders use advanced materials such as the carbon nanofibers now in laboratory study.

"His book brought the idea to the general public through the science fiction community," said Smitherman. But Clarke wasn't the first.

As early as 1895, a Russian scientist named Konstantin Tsiolkovsky suggested a fanciful "Celestial Castle" in geosynchronous Earth orbit attached to a tower on the ground, not unlike Paris's Eiffel tower. Another Russian, a Leningrad engineer by the name of Yuri Artsutanov, wrote some of the first modern ideas about space elevators in 1960. Published as a non-technical story in Pravda, his story never caught the attention of the West. Science magazine ran a short article in 1966 by John Isaacs, an American oceanographer, about a pair of whisker-thin wires extending to a geostationary satellite. The article ran basically unnoticed. The concept finally came to the attention of the space flight engineering community through a technical paper written in 1975 by Jerome Pearson of the Air Force Research Laboratory. This paper was the inspiration for Clarke's novel.

Pearson, who participated in the 1999 workshop, envisions the space elevator as a cost-cutting device for NASA. "One of the fundamental problems we face right now is that it's so unbelievably expensive to get things into orbit," said Pearson. "The space elevator may be the answer."The workshop's findings determined the energy required to move a payload by space elevator from the ground to geostationary orbit could remain relatively low. Using today's energy costs, researchers figured a 12,000-kg Space Shuttle payload would cost no more than $17,700 for an elevator trip to GEO. A passenger with baggage at 150 kg might cost only $222! "Compare that to today's cost of around $10,000 per pound ($22,000 per kg)," said Smitherman. "Potentially, we're talking about just a few dollars per kg with the elevator."

During the workshop, issues pertinent to transforming the concept from science fiction to reality were discussed in detail. "What the workshop found was there are real materials in laboratories today that may be strong enough to construct this type of system," said Smitherman.

Smitherman listed five primary technology thrusts as critical to the development of the elevator. First was the development of high-strength materials for both the cables (tethers) and the tower.

In a 1998 report, NASA applications of molecular nanotechnology, researchers noted that "maximum stress [on a space elevator cable] is at geosynchronous altitude so the cable must be thickest there and taper exponentially as it approaches Earth. Any potential material may be characterized by the taper factor--the ratio between the cable's radius at geosynchronous altitude and at the Earth's surface. For steel the taper factor is tens of thousands--clearly impossible. For diamond, the taper factor is 21.9 including a safety factor. Diamond is, however, brittle. Carbon nanotubes have a strength in tension similar to diamond, but bundles of these nanometer-scale radius tubes shouldn't propagate cracks nearly as well as the diamond tetrahedral lattice."

Fiber materials such as graphite, alumina, and quartz have exhibited tensile strengths greater than 20 GPa (Giga-Pascals, a unit of measurement for tensile strength) during laboratory testing for cable tethers. The desired strength for the space elevator is about 62 GPa. Carbon nanotubes have exceeded all other materials and appear to have a theoretical strength far above the desired range for space elevator structures. "The development of carbon nanotubes shows real promise," said Smitherman. "They're lightweight materials that are 100 times stronger than steel." The second technology thrust was the continuation of tether technology development to gain experience in the deployment and control of such long structures in space.

Third was the introduction of lightweight, composite structural materials to the general construction industry for the development of taller towers and buildings. "Buildings and towers can be constructed many kilometers high today using conventional construction materials and methods," said Smitherman. "There simply has not been a demonstrated need to do this that justifies the expense." Better materials may reduce the costs and make larger structures economical.

Fourth was the development of high-speed, electromagnetic propulsion for mass-transportation systems, launch systems, launch assist systems and high-velocity launch rails. These are, basically, higher speed versions of the trams now used at airports to carry passengers between terminals. They would float above the track, propelled by magnets, using no moving parts. This feature would allow the space elevator to attain high vehicle speeds without the wear and tear that wheeled vehicles would put on the structure.

Fifth was the development of transportation, utility and facility infrastructures to support space construction and industrial development from Earth out to GEO. The high cost of constructing a space elevator can only be justified by high usage, by both passengers and payload, tourists and space dwellers.

- Employing materials made from carbon nanotubes to reduce the weight of spaceships while retaining or even increasing the structural strength.

NANOTECHNOLOGY IN BATTERIES

How can Nanotechnology Improve Batteries?

Battery systems that fit in cars don't hold enough energy for driving distances, yet take hours to recharge and don't give much power for acceleration. Renewable sources like solar and wind deliver significant power only part time, but devices to store their energy are expensive and too inefficient to deliver enough power for surge demand.

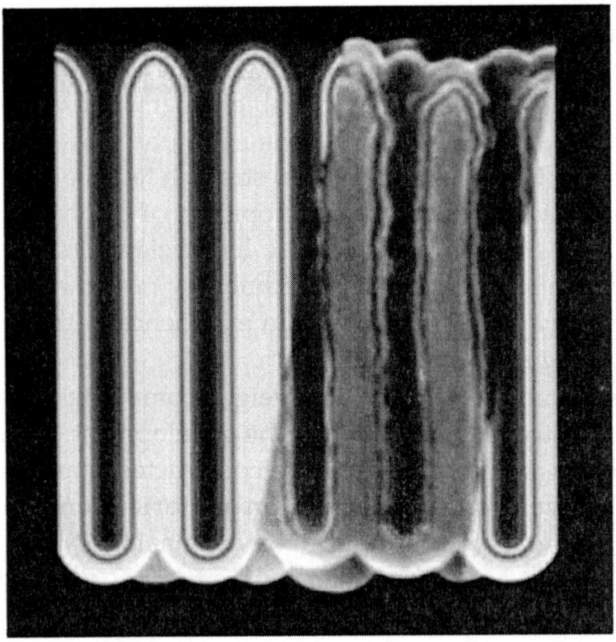

Figure: ***Nanotech Batteries***

Using nanotechnology in the manufacture of batteries offers the following benefits: Increasing the available power from a battery and decreasing the time required to recharge a battery. These benefits are achieved by coating the surface of an electrode with nanoparticles.

This increases the surface area of the electrode, thereby allowing more current to flow between the electrode and the chemicals inside the battery. This technique could increase the efficiency of hybrid vehicles by significantly reducing the weight of the batteries needed to provide adequate power. Traditional Li Ion technology uses active materials with particles that range in size between 5 and 20 microns.

These large particles are required to minimize safety risks inherent to first-generation Li Ion chemistries. A123 high power batteries are based on a safer and more stable active material that can use particle sizes below 100nm without adverse reaction. This new storage electrode enables much faster kinetics providing higher power than is possible from any other Li Ion chemistry.

Furthermore, to take advantage of the power delivered by this new chemistry, A123 has developed novel electrode and cell designs that provide the lowest impedance of any battery of its size, and a new electrolyte system that operates over a much wider temperature range. This new stable cell chemistry is environmentally friendly and inherently safer than other cathode chemistries.

Traditional Reserve Battery Technology

Many applications require a reserve battery. Typical implementations use a mechanical mechanism to combine the materials to create the battery. These technologies have a long shelf life, high reliability, high power densities, wide range of chemistries and wide operational range. But they suffer from:

- Cannot be easily miniaturized
- Incompatible with semiconductor processes (Cannot be integrated onto "chips".)
- Based upon mechanical activation, are slow to ramp up power
- Have to maintain separate electrolyte storage and activation mechanism taking up internal space
- Often are expensive.

Nanotechnology will Enable Unlimited shelf Life Batteries

A battery depends upon a chemical reaction. A variety of chemistries can be used. Nanotechnology provides a unique mechanism to combine the chemicals and control the characteristics of the reaction.

The image is a representative example of "nanograss" tubes. These tubes provide a "Superhydrophobic NanoStructured Surface" atop of which can be placed a droplet.

The droplet sits above the tubes with little or no interaction with the tubes themselves. But by careful engineering the droplet can be made to fall within the space between the tubes encountering a greatly increased surface area and interacting with the tubes themselves to causing current to flow.. The drop can be engineered to occur upon a variety of stimuli: voltage, RF and/or others. Advantages include:

- Long shelf life
- Easy to miniaturize
- Quick ramp up to full power
- Compatible with Semiconductor Processing
- High Power and Energy Density
- Wide range of chemistries can be used
- Inexpensive to mass produce
- Increasing the shelf life of a battery by using nanomaterials to separate liquids in the battery from the solid electrodes when there is no draw on the battery. This separation prevents the low level discharge that occurs in a conventional battery, which increases the shelf life of the battery dramatically.
- Reducing the possibility of batteries catching fire by providing less flammable electrode material.

NANOTECHNOLOGY IN SOLAR CELLS

How can Nanotech Improve Solar Cells?

Using nanoparticles in the manufacture of solar cells has the following benefits:

Reduced manufacturing costs as a result of using a low temperature process similar to printing instead of a high temperature vacuum deposition process typically used to produce conventional solar cells made with crystalline semiconductor material.

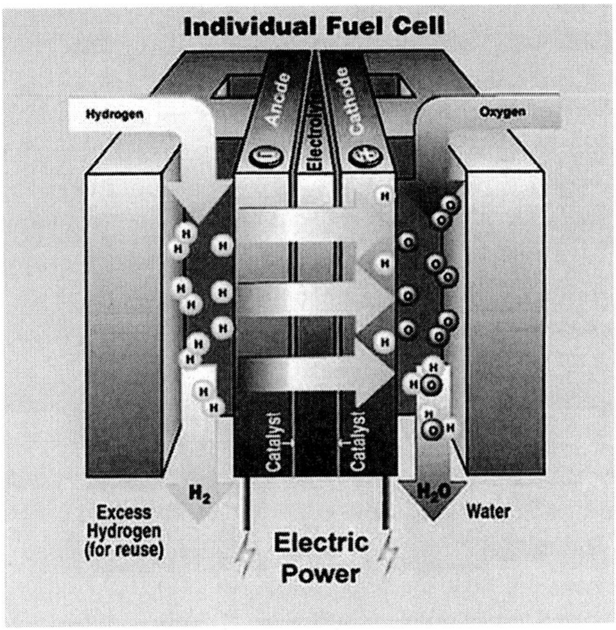

Figure: ***Individual Fuel Cell***

Reduced installation costs achieved by producing flexible rolls of solar cells instead of rigid crystalline panels. The new semiconductor thin film solar cells will also have this characteristic.

Currently available nanotechnology solar cells are not as efficient as traditional solar cells, however their lower cost offsets this. In the long term nanotechnology solar cells should both be lower cost and, using quantum dots, should be able to reach higher efficiency levels than conventional solar cells.

Nanotechnology Applications under Development

Nanoparticles in plastic film to form solar cells that can be incorporated into cases for devices such as cell phones and laptop computers. Semiconductor nanoparticles in a low temperature printing process that results in low cost thin film solar cells for large scale installations. The ability to architect and assemble materials on a nanometer scale now makes it possible to optimize solar cells at the very length scale at which the relevant photovoltaic semiconductor quantum-physics occurs. Molecular self-assembly techniques for instance now give us the unprecedented capability of designing and creating nanostructured materials with novel properties. Such techniques generally rely on formulas that control the precise, bottom-up chemical assembly of molecules into geometric structures composed of many molecules, e.g. in the 1nm to 100nm range.

We have developed a suite of in-house capabilities for creating nanostructured components based on various patented and patent-pending techniques. Nanosolar's component nanoengineering capabilities are focused on certain optoelectronically relevant materials and include semiconductor quantum dots and nanoparticles as well as nanotemplates with precise three-dimensional order. We use nanostructured components as the basis for creating printable semiconductors, printable transparent electrodes, novel forms of advanced nanocomposite solar-cell architecture, as well as powerful new forms of barrier films. We have developed a product and technology roadmap according to which an increasing range of such innovations can be incorporated into our products over time.

Organic molecules to make low cost solar cells.

Small Molecular Organic Photovoltaics-OPV

Photonic energy is the most abundant and available source of energy-yet remains virtually unused by mankind. The photovoltaic effect behind solar cells has been understood for years-first discovered in 1839. Yet the products developed to harvest solar energy remain fragile, heavy, expensive to manufacture and generally not competitive with other energy sources. Virtually all of the commercial products available today use inorganic materials, primarily silicon. Despite these challenges, the PV market has grown at over 20% annually for over 20 years-and recent growth has accelerated. In 2005, module production grew over 50% as sales reached $9 billion.

NANOTECHNOLOGICAL THRESHOLD

In this chapter, we outline how emerging technologies can lead to nanotechnology. The actual path to nanotechnology-the one that history books will record-could emerge from any one of the research directions in physics, biochemistry, and chemistry recounted in the last chapter, or (more likely) from a combination of them. The availability of so many good options builds confidence that the goal can be reached, even while it decreases confidence that some particular path will be fastest. To see how advances might cross the gap from present technology to early nanotechnology, let's follow one path out of the many possible.

One way to bridge the gap would through the development of an AFM-based molecular manipulator capable of doing primitive molecular manufacturing. This device would combine a simple molecular device-a molecular gripper-with an AFM positioning mechanism. An AFM can move its tip with precision; a molecular manipulator would add a gripper to the tip to hold a molecular tool. A molecular manipulator of this kind would guide chemical reactions by positioning molecules, like a slow, simple, but enormous assembler. (In our standard simulation view, where a molecular assembler arm fits in a room, the AFM apparatus of a molecular manipulator would be the size of a moon.) Despite its limits, an AFM molecular manipulator will be a striking advance.

How might this advance occur? Since we're choosing one path out of many possible, we may as well include more details and tell a story. (A more technical description of a device like the following can be found in Nature; see the technical bibliography).

Developing a Molecular Manipulator

Several years ago, researchers at the University of Brobdingnag began work on developing a molecular manipulator. To reach this goal, a team of a dozen physicists, chemists, and protein researchers banded together (some working full time, some part time) and began the creative teamwork needed to solve the basic problems.

First they needed to attach a gripper to an AFM tip. As grippers, they chose fragments of antibody molecules, the selectively sticky proteins that the immune system uses to bind and identify germs. If they could get the "back" of the molecule stuck onto a tip, then the "front" could bind and hold molecular tools. (The advantage of antibody fragments was this: freedom of tool choice. Since the late 1980s, researchers had been able to generate antibodies able to bind almost any preselected molecule-or molecular tool.)

A molecular manipulator would bind and position reactive molecular tools to build up a workpiece, molecule by molecule. In parallel, the U. Brob AFM researchers worked on placing tips in a precise location and then holding them there with atomic accuracy for seconds at a time. This proved straightforward. They used techniques developed elsewhere during the early 1990s, adding only modest refinements.

They now had their gripper and a way of putting it where they wanted it, but they needed a set of tools. The gripper was like the chuck of a drill, waiting to have different bits fitted into its tool-holder slot.

So as the final step, the synthetic chemists on the team made a dozen different molecular tools, all identical at one end but different at the other. The similar parts all bound to the same antibody tool-holder, slotting neatly into position. The different parts were all chemically reactive in different ways. Developing the molecular tool kit was the toughest part of the project; it took about as much work as had gone into duplicating the palytoxin molecule back in the 1980s. None of the tasks in the project demanded the solution of a deep scientific puzzle, and none demanded the solution of a notoriously difficult engineering problem. Each task had many possible solutions, the problem was to find a compatible set of solutions and apply them. After a few years, the solutions came together and the U. Brob research team began building new molecules by molecular manipulation. Now many teams are doing likewise.

Grippers and Tools

To build something with the U. Brob team's AFM-based molecular manipulator system, you use it as follows: First, choose a surface to build on and place it under the tip in a pool of liquid. Then dunk the AFM tip into the liquid, bringing it down to the surface, and back it off a little. Construction can now begin as soon as a tool is loaded into the gripper.

Tubes and pumps can flow different liquids over the surface and past the gripper, carrying different tool molecules. If you want to do something with a tool of Type A, you wash in the proper liquid.

Once it is in the gripper, you can use the AFM mechanism to move it around and put it where you want it. Move it up to the surface at a convenient spot, wait a few seconds, and it reacts, forming a bond and leaving a molecular fragment attached to the spot you chose. To add a different fragment, you can use a tool of Type B: you back up the tip, flow in a fresh liquid carrying the new tools, and in a moment a tool of the new type is bound in place and ready to apply, either on or alongside the first spot. Step by step, you build up a precise molecular structure.

Each step takes only seconds. Molecular tools pop into the gripper in a fraction of a second, and used tools pop off at the same rate. Once the tip has positioned a molecule, it reacts quickly, about a million times faster than unwanted reactions at other sites. In this way, the molecular manipulator gives good control of where reactions will occur (though it is not as reliable as an advanced assembler would be). It is fairly fast by a chemist's standards-per cycle-but still a million times slower than an advanced assembler. It can perform a variety of steps, but isn't as flexible and capable as an advanced assembler. In short, it is hardly the last word in nanotechnology, yet is a great advance over what has gone before.

Products

With its ability to accelerate desired reactions by a factor of a million or so, the U. Brob team's molecular manipulator can perform 10,000 to 100,000 steps with good reliability. Back in the 1980s, chemists making protein molecules struggled to perform just one hundred steps. The U. Brob research team (and its many imitators) can now build structures that are stronger and easier to design than proteins: not floppy, folded chains, but rugged objects held together by a sturdy network of bonds. Though not as strong and dense as diamond, these structures are like bits of a tough engineering plastic. A specially adapted computer-aided design system makes it easy to design molecular objects made from these materials.

Yet the AFM-based molecular manipulator has one grave disadvantage. It does chemistry one molecule at a time, and it ties up a machine as expensive as a car for hours or days to produce that one large molecule. Some molecules, though, are valuable enough to be worth building even one at a time. These draw prompt attention.

A single molecule isn't much use as a dye, a drug, or a floor wax, but it can have substantial value if it provides useful information. The U. Brob team quickly publishes a pile of scientific papers based on experiments with single molecules: they build a molecule, probe it, report the results, and build another. Some of these results show chemists elsewhere in the multibillion-dollar chemical industry how to design new catalysts, molecules that can help make other molecules more cheaply, cleanly, and efficiently. This information is worth a lot.

Three new products of special interest are among the first to be made. The first-molecular electronics-begins with experiments conducted by a research group at a computer chip company. They use their molecular manipulator to build single molecules and probe them, gradually learning how to build the parts needed for molecular electronic computers. These new computers don't immediately become practical, because the costs are too high for making such large molecules with AFM-based technology. Yet some companies begin to produce simpler molecular electronic devices for use in sensors and specialized high-speed signal processing. A specialty industry is born and begins to expand.

The second product is a gene reader, a complex molecular device built on the surface of a chip. The biologists who built the reader combined proteins borrowed from cells with special-purpose molecular machines designed from scratch. The result was a molecular system that binds DNA molecules and pulls them past a read-head-like tape through a tape recorder. The device works as fast as some naturally occurring molecular machines that read DNA, with one key advantage: it outputs its data electronically. At that speed, a single device can read a human genome in about a year. Though still too expensive for a doctor's office, these readers are promptly in great demand from research laboratories. Another small industry is born. The third product is far more important, in the long run: replacement tips for molecular manipulators, grippers, and tools that are better than the originals. With these new, more versatile devices, researchers are now building more ambitious products and tools.

MICROTECHNOLOGY LEADING TO NANOTECHNOLOGY

Microtechnology takes massive amounts of information and condenses it in a very small area, such as a silicon wafer or the microchip used in computers and mobile phones. The result: Countless new miniature products with complex engine rooms — medical devices, automobile computer systems, security and defense products like night-vision systems, airplane autopilot systems, logistics systems, even children's toys. Can bulldozers can be used to make wristwatches? At most, they can help to build factories in which watches are made. Though there could be surprises, the relevance of microtechnology to molecular nanotechnology seems similar. Instead, a bottom-up approach is needed to accomplish engineering goals on the molecular scale.

Figure: ***Microtechnology and Nanotechnology***

Tools Used for Molecular Engineering

Almost by definition, the path to molecular nanotechnology must lead through molecular engineering. Working in different disciplines, driven by different goals, researchers are making progress in this field. Chemists are developing techniques able to build precise molecular structures of sorts never before seen. Biochemists are learning to build structures of familiar kinds, such as proteins, to make new molecular objects.

In a visible sense, most of the tools used by chemists and biochemists are rather unimpressive. They work on countertops cluttered with dishes, bottles, tubes, and the like, mixing, stirring, heating, and pouring liquids-in biochemistry, the liquid is usually water with a trace of material dissolved in it. Periodically, a bit of liquid is put into a larger machine and a strip of paper comes out with a graph printed on it.

As one might guess from this description, research in the molecular sciences is usually much less expensive than research in high-energy physics (with its multibillion-dollar particle accelerators) or research in space (with its multibillion-dollar spacecraft). Chemistry has been called "small science," and not because of the size of the molecules. Of these fields, it is biomolecular science that is most obviously developing tools that can build nanotechnology, because biomolecules already form molecular machines, including devices resembling crude assemblers. This path is easiest to picture, and can surely work, yet there is no guarantee that it will be fastest: research groups following another path may well win. Each of these paths is being pursued worldwide, and on each, progress is accelerating.

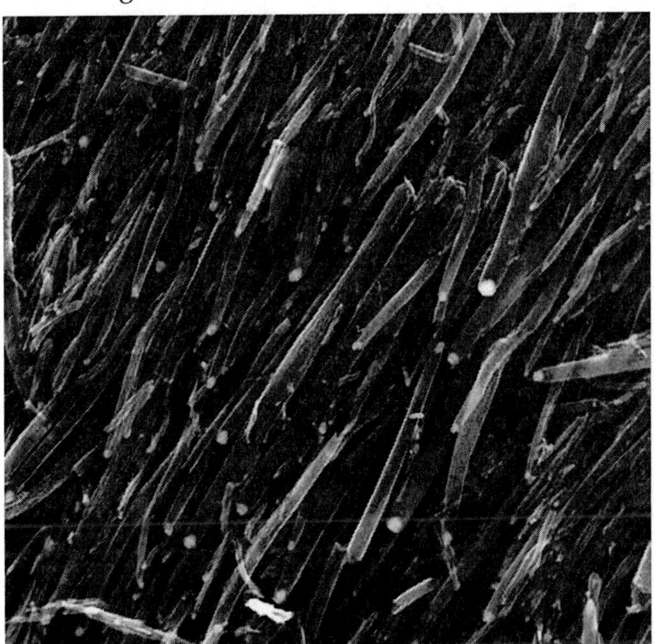

Figure: ***Therefore, nanotechnology is a subsection of microtechnology involving***

Physicists have recently contributed new tools of great promise for molecular engineering. These are the proximal probes, including the scanning tunneling microscope (STM) and the atomic force microscope (AFM). A proximal-probe device places a sharp tip in proximity to a surface and uses it to probe (and sometimes modify) the surface and any molecules that may be stuck to it.

MOLECULAR BUILDING BLOCKS

A huge technology base for molecular construction already exists. Tools originally developed by biochemists and biotechnologists to deal with molecular machines found in nature can be redirected to make new molecular machines. The expertise built up by chemists in more than a century of steady progress will be crucial in molecular design and construction. Both disciplines routinely handle molecules by the billions and get them to form patterns by self-assembly. Biochemists, in particular, can begin by copying designs from nature. Molecular building-block strategies could work together with proximal probe strategies, or could replace them, jumping directly to the construction of large numbers of molecular machines. Either way, protein molecules are likely to play a central role, as they do in nature.

Protein Engineering Building Molecular Machines

Proteins can self assemble into working molecular machines, objects that do something, such as cutting and splicing other molecules or making muscles contract. They also join with other molecules to form huge assemblies like the ribosome (about the size of a washing machine, in our simulation view). Ribosomes-programmable machines for manufacturing proteins-are nature's closest approach to a molecular assembler. The genetic-engineering industry is chiefly in the business of reprogramming natural nanomachines, the ribosomes, to make new proteins or to make familiar proteins more cheaply. Designing new proteins is termed protein engineering. Since biomolecules already form such complex devices, it's easy to see that advanced protein engineering could be used to build first-generation nanomachines.

FANCY MOLECULAR MACHINES

Making proteins is easier than designing them. Protein chemists began by studying proteins found in nature, but have only recently moved on to the problem of engineering new ones. These are called de novo proteins, meaning completely new, made from scratch. Designing proteins is difficult because of the way they are constructed. As Bill DeGrado, a protein chemist at Du Pont, explains: "A characteristic of proteins is that their activities depend on their three-dimensional structures. These activities may range from hormonal action to a function in digestion or in metabolism.

Whatever their function, it's always essential to have a definite three-dimensional shape or structure." This three-dimensional structure forms when a chain folds to form a compact molecular object. To get a feel for how tough it is to predict the natural folding of a protein chain, picture a straight piece of cord with hundreds of magnets and sticky knots along its length. In this state, it's easy to make and easy to understand. Now pick it up, put it in a glass jar, and shake it for a long time. Could you predict its final shape? Certainly not: it's a tangled mess. One might call this effort at prediction "the sticky-cord-folding problem"; protein chemists call theirs "the protein-folding problem."

Given the correct conditions, a protein chain always folds into one special shape, but that shape is hard to predict from just the straightened structure. Protein designers, though, face the different job of first determining a desired final shape, and then figuring out what linear sequence of amino acids to use to make that shape. Without solving the classic protein-folding problem, they have begun to solve the protein-design problem.

THE ACCOMPLISHMENTS SO FAR

Bill DeGrado and his colleagues at Du Pont had one of the first successes: "We've been able to use basic principles to design and build a simple molecule that folds up the way we want it to. This is really the first real example of a designed protein structure, designed from scratch, not by taking an already existing structure and tinkering with it."

Although scientists do the work, the work itself is really a form of engineering, as shown by the title of the field's journal, Protein Engineering. Bill DeGrado's description of the process makes this clear: "After you've made it, the next step is to find out whether your protein did what you expected it to do. Did it fold? Did it pass ions across bilayers [such as cell membranes]? Does it have a catalytic function [speeding specific chemical reactions]? And that's tested using the appropriate experiment. More than likely, it won't have done what you wanted it to do, so you have to find out why. Now, a good design has in it a contingency plan for failure and helps you learn from mistakes. Rather than designing a structure that would take a year or more to analyze, you design it so that it can be assayed for given function or structure in a matter of days."

Many groups are pursuing protein design today, including academic researchers like Jane and Dave Richardson at Duke University, Bruce Erickson at the University of North Carolina, and Tom Blundell, Robin Leatherbarrow, and Alan Fersht in Britain. The successes have started to roll in. Japan, however, is unique in having an organization devoted exclusively to such projects: the Protein Engineering Research Institute (PERI) in Osaka. In 1990, PERI announced the successful design and construction of a de novo protein several times larger than any built before.

Speciality of Proteins

The main advantage of proteins is that they are familiar: a lot is known about them, and many tools exist for working with them. Yet proteins have disadvantages as well. Just because this design work is starting with proteins-soft, squishy molecules that are only marginally suitable for nanotechnology-doesn't mean it will stay within those limits.

De Grado points out "The fundamental goal of our work in de novo design is to be able to take the next step and get entirely away from protein systems." An early example is the work of Wallace Carothers of Du Pont, who used a de novo approach to studying the nature of proteins: Rather than trying to cut up proteins, he tried to build up things starting with amino acids and other similar monomers. In 1935, he succeeded in making nylon.

DeGrado explains "There is a deep philosophical belief at Du Pont in the ability of people to make molecules de novo that will do useful things. And there is a fair degree of commitment from the management that following that path will lead to products: not directly, and not always predictably, but they know that they need to support the basic science. "I think ultimately we have a better chance at doing some really exciting things by de novo design, because our repertory should be much greater than that of nature. Think about the ability to fly: One could breed better carrier pigeons or one could design airplanes." The biology community, however, leans more toward ornithology than toward aerospace engineering. DeGrado's experience is that "a lot of biologists feel that if you aren't working with the real thing [natural proteins], you aren't studying biology, so they don't totally accept what we're doing. On the other hand, they recognize it as good chemistry."

Where is Protein Engineering Headed?

Like the IBM physicists, protein designers are moved by a vision of molecular engineering. In 1989, Bill DeGrado predicted, "I think we'll be able to make catalysts or enzymelike molecules, possibly ones that catalyze reactions not catalyzed in nature." Catalysts are molecular machines that speed up chemical reactions: they form a shape for the two reacting molecules to fit into and thereby help the reaction move faster, up to a million reactions per second. New ones, for reactions that now go slowly, will give enormous cost savings to the chemical industry.

This prediction was borne out just a few months later, when Denver researchers John Stewart, Karl Hahn, and Wieslaw Klis announced their new enzyme, designed from scratch over a period of two years and built successfully on the first try.

It's a catalyst, making some reactions go about 100,000 times faster. Nobel Prize-winning biochemist Bruce Merrifield believes that "if others can reproduce and expand on this work, it will be one of the most important achievements in biology or chemistry."

DeGrado also has longer term plans for protein design, beyond making new catalysts: "It will allow us to think about designing molecular devices in the next five to ten years. It should be possible ultimately to specify a particular design and build it.

Then you'll have, say, proteinlike molecules that self-assemble into complex molecular objects, which can serve as machinery. But there's a limit to how small you can make devices. You'll shrink things down so far and then you won't be able to go any further, because you've reached molecular dimensions."

Mark Pearson shows that management at Du Pont also has this vision. Regarding the prospects for nanotechnology and assemblers, he remarked, "You know, it'll take money and effort and good ideas for sure. But to my way of thinking, there is no absolute fundamental limitation to preclude us from doing this kind of thing."

He didn't say his company plans to develop nanotechnology, but such plans aren't really necessary. Du Pont is already on the nanotechnology path, for other-shorter-term, commercial-reasons. Like IBM, if they do decide to move quickly, they have the resources and forward-looking people needed to succeed.

Who Else Builds Molecular Objects?

Chemists, most of whom do not work on proteins, are the traditional experts in building molecular objects. As a group they've been building molecules for over a century, with ever increasing ability and confidence. Their methods are all indirect: They work with billions of atoms at a time-massive parallelism-but without control of the positions of their workpieces.

The molecules typically tumble randomly in a liquid or gas, like pieces of a puzzle that may or may not fit together correctly when shaken together in a box. With clever design and planning, most pieces will join properly.

Chemists mix molecules on a huge scale (in our simulation view, a test tube holds a churning molecular swarm with the volume of an inland sea), yet they still achieve precise molecular transformations. Given that they work so indirectly, their achievements are astounding.

This is, in part, the result of the enormous amount of work poured into the field for many decades. Thousands of chemists are working on molecular construction in the United States alone; add to that the chemists in Europe, in Japan, and in the rest of the world, and you have a huge community of researchers making great strides.

Though it publishes only a one-paragraph summary of each research report, a guide to the chemical literature-Chemical Abstracts-covers several library walls and grows by many feet of shelf space every year.

How can Mixing Chemicals Build Molecular Objects?

An engineer would say that chemists (at least those specializing in synthesis) are doing construction work, and would be amazed that they can accomplish anything without being able to grab parts and put them in place. Chemists, in effect, work with their hands tied behind their backs.

Molecular manufacturing can be termed "positional chemistry" or "positional synthesis," and will give chemists the ability to put molecules where they want them in three-dimensional space.

Rather than trying to design puzzle pieces that will stick together properly by themselves when shaken together in a box, chemists will then be able to treat molecules more like bricks to be stacked. The basic principles of chemistry will be the same, but strategies for construction will become far simpler.

Without positional control, chemists face a problem something like this: Picture a giant glass barrel full of tiny battery-powered drills, buzzing away in all directions, vibrating around in the barrel. Your goal is to take a piece of wood and put a hole in just one specific spot. If you simply throw it in the barrel, it will be drilled haphazardly in many places.

To control the process, you must protect all the places you don't want drilled-perhaps by gluing protective pieces of metal over most of the wood surface. This problem-how to protect one part of a molecule while altering another part-has forced chemists to develop ever-cleverer ploys to build larger and larger molecules.

Fancy Molecular Machines

Chemists can achieve great things, but have focused much of their effort on duplicating molecules found in nature and then making minor variants. As an example, take palytoxin, a molecule found in a Hawaiian coral. It was so difficult to make in the lab that it has been called "the Mount Everest of synthetic chemistry," and its synthesis was hailed as a triumph.

Other efforts are poured into making small molecules with unusual bonding, or molecules of remarkable symmetry, like "cubane" and "dodecahedrane" (shaped like the Platonic solids they are named after).

Chemists, at least in the United States, regard themselves as natural scientists even when their life's work is the construction of molecules by artificial means.

Ordinarily, people who build things are called engineers. And indeed, at the University of Tokyo the Department of Synthetic Chemistry is part of the Faculty of Engineering; its chemists are designing molecular switches for storing computer data. Engineering achievements will require work directed at engineering goals.

Building Molecular Machines

Molecular engineers working toward nanotechnology need a set of molecular building blocks for making large, complex structures. Systematic building-block construction was pioneered by Bruce Merrifield, winner of the 1984 Nobel Prize in Chemistry. His approach, known as "solid phase synthesis," or simply "the Merrifield method," is used to synthesize the long chains of amino acids that form proteins. In the Merrifield method, cycles of chemical reaction each add one molecular building block to the end of a chain anchored to a solid support. This happens in parallel to each of trillions of identical chains, building up trillions of molecular objects with a particular sequence of building blocks. Chemists routinely use the Merrifield method to make molecules larger than palytoxin, and related techniques are used for making DNA in so-called gene machines: an ad from an Alabama company reads, "Custom DNA-Purified and Delivered in 48 hours."

While it's hard to predict how a natural protein chain will fold-they weren't designed to fold predictably-chemists could make building blocks that are larger, more diverse, and more inclined to fold up in a single, obvious, stable pattern. With a set of building blocks like these, and the Merrifield method to string them together, molecular engineers could design and build molecular machines with greater ease.

How do Researchers Design what they can't See?

To make a new molecule, both its structure and the procedure to make it must be designed. Compared to gigantic science projects like the Superconducting Supercollider and the Hubble Space Telescope, working with molecules can be done on a shoestring budget. Still, the costs of trying many different procedures add up. To help predict in advance what will work and what won't, designers turn to models.

You may have played with molecular models in chemistry class: colored plastic balls and sticks that fit together like Tinker Toys. Each color represents a different kind of atom: carbon, hydrogen, and so on. Even simple plastic models can give you a feel for how many bonds each kind of atom makes, how long the bonds are, and at what angles they are made. A more sophisticated form of model uses only spheres and partial spheres, without sticks. These colorful, bumpy shapes are called CPK models, and are widely used by professional chemists. Nobel laureate Donald Cram remarks that "We have spent hundreds of hours building CPK models of potential complexes and grading them for desirability as research targets." His research, like that of fellow Nobelists Charles J. Pedersen and Jean-Marie Lehn, has focused on designing and making medium-sized molecules that self assemble.

Although physical models can't give a good description of how molecules bend and move, computer-based molecules can. Computer-based modeling is already playing a key role in molecular engineering. As John Walker (a founder and leader of Autodesk) has remarked, "Unlike all of the industrial revolutions that preceded it, molecular engineering requires, as an essential component, the ability to design, model, and simulate molecular structures using computers." This has not gone unnoticed in the business community. John Walker's remark was part of a talk on nanotechnology given at Autodesk, a leader in computer-aided design and one of the five largest software firms in the United States. Soon after this talk, the company made its first major investment in the computer-aided design of molecules.

DECISIONS AFFECTING THE RATE OF ADVANCE

Decisions about research directions are central. Researchers are already pouring effort into chemical synthesis, molecular engineering, and related fields. The same amount of effort could produce more impressive results in molecular nanotechnology if a fraction of it were differently directed. The research funders-corporate executives, and decision makers in science funding agencies like the National Science Foundation in the United States and Japan's Ministry of International Trade and Industry-all have a large influence on research directions, but so do the researchers working in the labs. They submit proposals to potential funders (and often spend time on personally chosen projects, regardless of funding), so their opinions also shape what happens. Where public money is involved, politicians' impressions of public opinion can have a huge influence, and public opinion depends on what all of us think and say..

Still, researchers play a central role. They tend to work on what they think is interesting, which depends on what they think is possible, which depends on the tools they have or-among the most creative researchers-on the tools they can see how to make. Our tools shape how we think: as the saying goes, when all you have is a hammer, everything looks like a nail. New tools encourage new thoughts and enable new achievements, and decisions about tool development will pace advances in nanotechnology. To understand the challenges ahead, we need to take a look at ideas about the tools that will be needed.

IMPORATANCE OF TOOLS

Throughout history, limited tools have limited achievement. Leonardo da Vinci's sixteenth century chain drives and ball bearings were theoretically workable, yet never worked in their inventor's lifetime. Charles Babbage's nineteenth century mechanical computer suffered the same fate. The problem?

Both inventors needed precisely machined parts that (though readily available today) were beyond the manufacturing technology of their times. Physicist David Miller recounts how a sophisticated integrated circuit design project at TRW hit similar limits in the early 1980s: "It all came down to whether a German company could cool their glass lenses slowly enough to give us the accuracy we needed. They couldn't."

In the molecular world, tool development again paces progress, and new tools can bring breathtaking advances. Mark Pearson, director of molecular biology for Du Pont, has observed this in action: "When I was a graduate student back in the 1950s, it was a multiyear problem to determine the molecular structure of a single protein. We used to say, 'one protein, one career.' Yet now the time has shrunk from a career to a decade to a year-and in optimal cases to a few months." Protein structures can be mapped atom by atom by studying X-ray reflections from layers in protein crystals. Pearson observes that "Characterizing a protein was a career-long endeavor in part because it was so difficult to get crystals, and just getting the material was a big constraint. With new technologies, we can get our hands on the material now-that may sound mundane, but it's a great advance. To the people in the field, it makes all the difference in the world." Improved tools for making and studying proteins are of special importance because proteins are promising building blocks for first-generation molecular machines.

Science Discovery

Nobel Prizes are more often awarded for discoveries than for the tools (including instruments and techniques) that made them possible. If the goal is to spur scientific progress, this is a shame. This pattern of reward extends throughout science, leading to a chronic underinvestment in developing new tools. Philip Abelson, an editor of the journal Science, points out that the United States suffers from "a lack of support for development of new instrumentation. At one time, we had a virtual monopoly in pioneering advances in instrumentation. Now practically no federal funds are available to universities for the purpose." It's easier and less risky to squeeze one more piece of data out of an existing tool than to pioneer the development of a new one, and it takes less imagination.

But new tools emerge anyway, often from sources in other fields. The study of protein crystals, for example, can benefit from new X-ray sources developed by physicists, and techniques from chemistry can help make new proteins. Because they can't anticipate tools resulting from innovations in other fields, scientists and engineers are often too pessimistic about what can be achieved in their own fields. Nanotechnology will join several fields, and yield tools useful in many others. We should expect surprising results.

RESEARCHERS' TOOLS

Today's tools for making small-scale structures are of two kinds: molecular-processing tools and bulk-processing tools. For decades, chemists and molecular biologists have been using better and better molecular-processing tools to make and manipulate precise, molecular structures. These tools are of obvious use. Physicists, as we will see, have recently developed tools that can also manipulate molecules.

Combined with techniques from chemistry and molecular biology, these physicist's tools promise great advances. Microtechnologists have applied chip-making techniques to the manufacture of microscopic machines. These technologies-the main approach to miniaturization in recent decades-can play at most a supporting role in the development of nanotechnology. Despite appearances, it seems that microtechnology cannot be refined into nanotechnology.

SMALL MICROTECHNOLOGY

For many years, it was conventional to assume that the road to very small devices led through smaller and smaller devices: a top-down path. On this path, progress is measured by miniaturization: How small a transistor can we build?

How small a motor? How thin a line can we draw on the surface of a crystal? Miniaturization focuses on scale and has paid off well, spawning industries ranging from watchmaking to microelectronics.

Researchers at AT&T Bell Labs, the University of California at Berkeley, and other laboratories in the United States have used micromachining (based on microelectronic technologies) to make tiny gears and even electric motors. Micromachining is also being pursued successfully in Japan and Germany.

These microgears and micromotors are, however, enormous by nanotechnological standards: a typical device is measured in tens of micrometers, billions of times the volume of comparable nanogears and nanomotors. (In our simulated molecular world, ten microns is the size of a small town.) In size, confusing microtechnology with molecular nanotechnology is like confusing an elephant with a ladybug.

The differences run deeper, though. Microtechnology dumps atoms on surfaces and digs them away again in bulk, with no regard for which atom goes where. Its methods are inherently crude. Molecular nanotechnology, in contrast, positions each atom with care. As Bill DeGrado, a protein chemist at Du Pont, says, "The essence of nanotechnology is that people have worked for years making things smaller and smaller until we're approaching molecular dimensions. At that point, one can't make smaller things except by starting with molecules and building them up into assemblies." The difference is basic: In microtechnology, the challenge is to build smaller; in nanotechnology, the challenge is to build bigger-we can already make small molecules.

A language warning: in recent years, nanotechnology has indeed been used to mean "very small microtechnology"; for this usage, the answer to the above question is yes, by definition. This use of a new word for a mere extension of an old technology will produce considerable confusion, particularly in light of the widespread use of nanotechnology in the sense found here. Nanolithography, nanoelectronics, nanocomposites, nanofabrication: not all that is nano- is molecular, or very relevant to the concerns raised in this book. The terms molecular nanotechnology and molecular manufacturing are more awkward but avoid this confusion.

2

Nanomaterials

Functional Nanomaterials is the first and unique compilation of the state-of-the-art review chapters covering all aspects of functional nanomaterials and their applications. Nanotechnology has led to a profound paradigm shift after the developments in recent years and after being classified as one of the most important areas of impending technology by the U.S. government. Novel functional nanomaterials are the basis of newly emerging nanotechnologies for various device applications. This book with 30 chapters reflects the tremendous world-wide interest in functional nanostructured materials. The wide variety of topics covered in this book is interesting for professionals working in the fundamental and applied research. The book covers major classes of nanomaterials such as carbon nanotubes, carbon and polymer nano fibers, nano particles, nanocomposites, nanosheets, fullerenes, supramolecular and self-assembled nano structures, and many other types of nanomaterials. In addition, this book highlights various physico-chemical properties of different nanostructures such as catalytic, dielectric, magnetic, fluorescent and luminescent properties as well as applications of nanomaterials in sensors, photonics, drug delivery, proteomics, biomolecular electronics, and Homeland Security.

Much of nanoscience and many nanotechnologies are concerned with producing new or enhanced materials. Nanomaterials can be constructed by 'top down' techniques, producing very small structures from larger pieces of material, for example by etching to create circuits on the surface of a silicon microchip. They may also be constructed by 'bottom up' techniques, atom by atom or molecule by molecule. One way of doing this is self-assembly, in which the atoms or molecules arrange themselves into a structure due to their natural properties. Crystals grown for the semiconductor industry provide an example of self assembly, as does chemical synthesis of large molecules. A second way is to use tools to move each atom or molecule individually. Although this 'positional assembly' offers greater control over construction, it is currently very laborious and not suitable for industrial applications. Current applications of nanoscale materials include very thin coatings used, for example, in electronics and active surfaces (for example, self-cleaning windows). In most applications the nanoscale components will be fixed or embedded but in some, such as those used in cosmetics and in some pilot environmental remediation applications, free nanoparticles are used. The ability to machine materials to very high precision and accuracy (better than 100nm) is leading to considerable benefits in a wide range of industrial sectors, for example in the production of components for the information and communication technology, automotive and aerospace industries.

NANOMATERIALS- DEFINITION

Although a broad definition, we categorise nanomaterials as those which have structured components with at least one dimension less than 100nm. Materials that have one dimension in the nanoscale (and are extended in the other two dimensions) are layers, such as a thin films or surface coatings.

Figure: ***Future technology, nanomaterials, nanotechnology***

Some of the features on computer chips come in this category. Materials that are nanoscale in two dimensions (and extended in one dimension) include nanowires and nanotubes. Materials that are nanoscale in three dimensions are particles, for example precipitates, colloids and quantum dots (tiny particles of semiconductor materials). Nanocrystalline materials, made up of nanometre-sized grains, also fall into this category. Some of these materials have been available for some time; others are genuinely new. The aim of this chapter is to give an overview of the properties, and the significant foreseeable applications of some key nanomaterials.

Two principal factors cause the properties of nanomaterials to differ significantly from other materials: increased relative surface area, and quantum effects. These factors can change or enhance properties such as reactivity, strength and electrical characteristics. As a particle decreases in size, a greater proportion of atoms are found at the surface compared to those inside. For example, a particle of size 30 nm has 5% of its atoms on its surface, at 10 nm 20% of its atoms, and at 3 nm 50% of its atoms. Thus nanoparticles have a much greater surface area per unit mass compared with larger particles. As growth and catalytic chemical reactions occur at surfaces, this means that a given mass of material in nanoparticulate form will be much more reactive than the same mass of material made up of larger particles.

To understand the effect of particle size on surface area, consider a U.S. silver dollar. The silver dollar contains 26.96 grams of coin silver, has a diameter of about 40 mm, and has a total surface area of approximately 27.70 square centimeters. If the same amount of coin silver were divided into tiny particles – say 1 nanometer in diameter – the total surface area of those particles would be 11,400 square meters. When the amount of coin silver contained in a silver dollar is rendered into 1 nm particles, the surface area of those particles is 4.115 million times greater than the surface area of the silver dollar!

NANOMATERIAL TECHNOLOGY

Nanomaterial-based catalysts are usually heterogeneous catalysts broken up into nanoparticles in order to speed up the catalytic process. The extremely small size of the particles maximizes the surface area exposed to the reactant, allowing more reactions to occur at the same time, thus speeding up the process. Much research on nanomaterial-based catalysts have to do with maximizing the effectiveness of the catalyst coating in fuel cells. Platinum is currently the most common catalyst for this application, however, it is expensive and rare, so a lot of research has been going into maximizing the catalytic properties of other metals by shrinking them to nanoparticles in the hope that someday they will be an efficient and economic alternative to platinum.

Defense programs in many countries are now concentrating on nanotechnology research that will facilitate advances in such technology used to create secure but small messaging equipment, allow the development of smart weapons, improve stealth capabilities, aid in developing specialized sensors (including bio-inclusive sensors), help to create self-repairing military equipment, and improve the development and delivery mechanisms for medicines and vaccines. Nanotechnology builds on advances in microelectronics during the last decades of the twentieth century. The miniaturization of electrical components greatly increased the utility and portability of computers, imaging equipment, microphones, and other electronics. Indeed, the production and wide use of such commonplace devices such as personal computers and cell phones was absolutely dependent on advances in microtechnology.

Despite these fundamental advances there remain real physical constraints (e.g., microchip design limitations) to further miniaturization based upon conventional engineering principles. Nanotechnologies intend to revolutionize components and manufacturing techniques to overcome these fundamental limitations. In addition, there are classes of biosensors and feedback control devices that require nanotechnology because-despite advances in microtechnology-present components remain too large or slow.

ADVANCES

Nanotechnology advances affect all branches of engineering and science that deal directly with device components ranging in size between 1/10,000,000 (one ten millionth of a millimeter) and 1/10,0000 millimeter. At these scales, even the most sophisticated microtechnology-based instrumentation is useless. Engineers anticipate that advances in nanotechnology will allow the direct manipulation of molecules in biological samples (e.g., proteins or nucleic acids) paving the way for the development of new materials that have a biological component or that can provide a biological interface.

In addition to new tools, nanotechnology programs advance practical understanding of quantum physics. The internalization of quantum concepts is a necessary component of nanotechnology research programs because the laws of classical physics (e.g., classical mechanics or generalized gas laws) do not always apply to the atomic and near-atomic level. Nanotechnology and quantum physics. Quantum theory and mechanics describe the relationship between energy and matter on the atomic and subatomic scale. At the beginning of the twentieth century, German physicist Maxwell Planck (1858-1947) proposed that atoms absorb or emit electromagnetic radiation in bundles of energy termed quanta. This quantum concept seemed counter-intuitive to well-established Newtonian physics. Advancements associated with quantum mechanics (e.g., the uncertainty principle) also had profound implications with regard to the philosophical scientific arguments regarding the limitations of human knowledge.

Planck's quantum theory, which also asserted that the energy of light (a photon) was directly proportional to its frequency, proved a powerful concept that accounted for a wide range of physical phenomena. Planck's constant relates the energy of a photon with the frequency of light. Along with the constant for the speed of light, Planck's constant (h = 6.626 x 10-34 Joule-second) is a fundamental constant of nature.

Prior to Planck's work, electromagnetic radiation (light) was thought to travel in waves with an infinite number of available frequencies and wavelengths. Planck's work focused on attempting to explain the limited spectrum of light emitted by hot objects. Danish physicist Niels Bohr (1885-1962) studied Planck's quantum theory of radiation and worked in England with physicists J. J. Thomson (1856-1940), and Ernest Rutherford (1871-1937) to improve their classical models of the atom by incorporating quantum theory. During this time, Bohr developed his model of atomic structure.

According to the Bohr model, when an electron is excited by energy it jumps from its ground state to an excited state (i.e., a higher energy orbital). The excited atom can then emit energy only in certain (quantized) amounts as its electrons jump back to lower energy orbits located closer to the nucleus. This excess energy is emitted in quanta of electromagnetic radiation (photons of light) that have exactly the same energy as the difference in energy between the orbits jumped by the electron.

The electron quantum leaps between orbits proposed by the Bohr model accounted for Plank's observations that atoms emit or absorb electromagnetic radiation in quanta. Bohr's model also explained many important properties of the photoelectric effect described by Albert Einstein (1879-1955). Einstein assumed that light was transmitted as a stream of particles termed photons. By extending the well-known wave properties of light to include a treatment of light as a stream of photons, Einstein was able to explain the photoelectric effect. Photoelectric properties are key to regulation of many microtechnology and proposed nanotechnology level systems.

Quantum mechanics ultimately replaced electron "orbitals" of earlier atomic models with allowable values for angular momentum (angular velocity multiplied by mass) and depicted electron positions in terms of probability "clouds" and regions.

In the 1920s, the concept of quantization and its application to physical phenomena was further advanced by more mathematically complex models based on the work of the French physicist Louis Victor de Broglie (1892-1987) and Austrian physicist Erwin Schrödinger (1887-1961) that depicted the particle and wave nature of electrons. De Broglie showed that the electron was not merely a particle but a waveform. This proposal led Schrödinger to publish his wave equation in 1926. Schrödinger's work described electrons as a "standing wave" surrounding the nucleus, and his system of quantum mechanics is called wave mechanics. German physicist Max Born (1882-1970) and English physicist

P. A. M. Dirac (1902-1984) made further advances in defining the subatomic particles (principally the electron) as a wave rather than as a particle and in reconciling portions of quantum theory with relativity theory.

Working at about the same time, German physicist Werner Heisenberg (1901-1976) formulated the first complete and self-consistent theory of quantum mechanics.

Matrix mathematics was well established by the 1920s, and Heisenberg applied this powerful tool to quantum mechanics. In 1926, Heisenberg put forward his uncertainty principle which states that two complementary properties of a system, such as position and momentum, can never both be known exactly.

This proposition helped cement the dual nature of particles (e.g., light can be described as having both wave and particle characteristics). Electromagnetic radiation (one region of the spectrum that comprises visible light) is now understood to have both particle and wave like properties.

In 1925, Austrian-born physicist Wolfgang Pauli (1900-1958) published the Pauli exclusion principle states that no two electrons in an atom can simultaneously occupy the same quantum state (i.e., energy state). Pauli's specification of spin (+1/2 or-1/2) on an electron gave the two electrons in any suborbital differing quantum numbers (a system used to describe the quantum state) and made completely understandable the structure of the periodic table in terms of electron configurations (i.e., the energy-related arrangement of electrons in energy shells and suborbitals).

In 1931, American chemist Linus Pauling published a paper that used quantum mechanics to explain how two electrons, from two different atoms, are shared to make a covalent bond between the two atoms. Pauling's work provided the connection needed in order to fully apply the new quantum theory to chemical reactions.

Advances in nanotechnology depend upon an understanding and application of these fundamental quantum principles. At the quantum level the smoothness of classical physics disappears and nanotechnologies are predicated on exploiting this quantum roughness.

NANOTECHNOLOGY RESEARCH

In the United States, expenditures on nanotechnology development tops $500 million per year and is largely coordinated by the National Science Foundation and Department of Defense Advanced Research Projects Agency (DARPA) under the umbrella of the National Nano-technology Initiative. Other institutions with dedicated funding for nanotechnology include the Department of Energy (DOE) and National Institutes of Health (NIH).

Figure: ***Saudi Arabia boosts nanotech research***

Research interests. Current research interests in nano-technology include programs to develop and exploit nanotubes for their ability to provide extremely strong bonds. Nanotubes can be flexed and woven into fibers for use in ultrastrong-but also ultralight-bulletproof vests. Nanotubes are also excellent conductors that can be used to develop precise electronic circuitry.

Other interests include the development of nanotechnology-based sensors that allow smarter autonomous weapons capable of a greater range of adaptations enroute to a target; materials that offer stealth characteristics across a broader span of the electromagnetic spectrum; self-repairing structures; and nanotechnology-based weapons to disrupt-but not destroy-electrical system infrastructure.

A CYCLE OF FALLING COST

The above estimate made a conservative assumption about future costs: that energy and materials will cost then what they do now, before molecular manufacturing has become available. They won't, because lower costs lead to lower costs. Let's say that making one kilogram of product by molecular manufacturing requires one dollar for a kilogram of raw materials and four dollars for a generous forty kilowatt-hours of energy. These are typical present-day prices for materials and electrical energy. Assume, for the moment, that other costs are small. One of the resulting five-dollar-per-kilogram products can be solar cell paint suitable for applying to paved roads. A layer of paint a few millionths of a meter thick would cost about five cents per square meter to produce, and would generate enough energy to make another square meter of paint in less than a week, even allowing for nighttime and moderate cloud cover. The so-called energy payback time would thus be short.

Let's assume that this smart paint costs as much to spread and hook up as it does to make, and that we demand that it pay for itself in a single month, so we charge ten cents per square meter per month. At that rate, the cost of solar energy from resurfaced roads would be roughly $0.004 per kilowatt hour-less than a twentieth the energy cost assumed in the initial production-cost estimate. By itself, this makes the cost of production fall to a fraction of what it was before. Most of that remaining fraction consists of the cost of materials.

But the products of nanotechnology will mostly be made of carbon (if present expectations are any guide), and carbon dioxide is too abundant in the atmosphere these days. With energy so cheap, the atmosphere can be used as source of carbon (and of hydrogen, nitrogen, and oxygen). The price of carbon would be a few cents per kilogram-roughly a twentieth the original price assumed for raw materials.

But now, both energy and raw materials are a twentieth the original price, and so the products become cheaper, including the energy-producing products and the raw-material-producing (atmosphere-cleaning) products....

The above scenario is simple, but it seems realistic in its basic outlines: lower costs can lead to lower costs. How far this process can go is hard to estimate precisely, but it could go far indeed.

TOO CHEAP POWER

This argument will remind some readers of an old claim-that nuclear energy would lead to "power too cheap to meter." This assertion, attributed to the early nuclear era, has passed into folklore as a warning to be skeptical of technologists promising free goodies. Does the warning apply here?

Anyone claiming that something is free doesn't really understand economics. Using something always has a cost equal to the most valuable alternative use for the thing. Choosing one alternative sacrifices another, and that sacrifice is the cost. As economist Phillip K. Salin says, "There's no such thing as a free opportunity," since opportunities always cost (at least) time and attention. Nanotechnology will not mean free goodies. But, one might argue, nuclear power hasn't even been inexpensive. If technologists could be so wrong back then, why believe a similar argument today? We are happy to report that the arguments aren't similar: any argument for "nuclear power too cheap to meter" had to be absurd even given the knowledge at the time, and our argument isn't.

Nuclear reactors boil water to make steam to turn turbines to turn generators to drive electrical power through power lines to transformers to local power lines to houses, factories, and so forth. The wildest optimist could never have claimed that nuclear power was a free source of anything more than heat, and a realist would have added in the cost of the reactor equipment, fuel, waste disposal, hazards, and the rest. Even our wild optimist would have had to include the cost of building the boiler, the turbines, the generators, the power lines, and the transformers, and the cost of maintenance on all these. These costs were known to be a major part of the cost of power, so free heat wouldn't have meant free power. Thus, the claim was absurd the day it was made-not merely in hindsight.

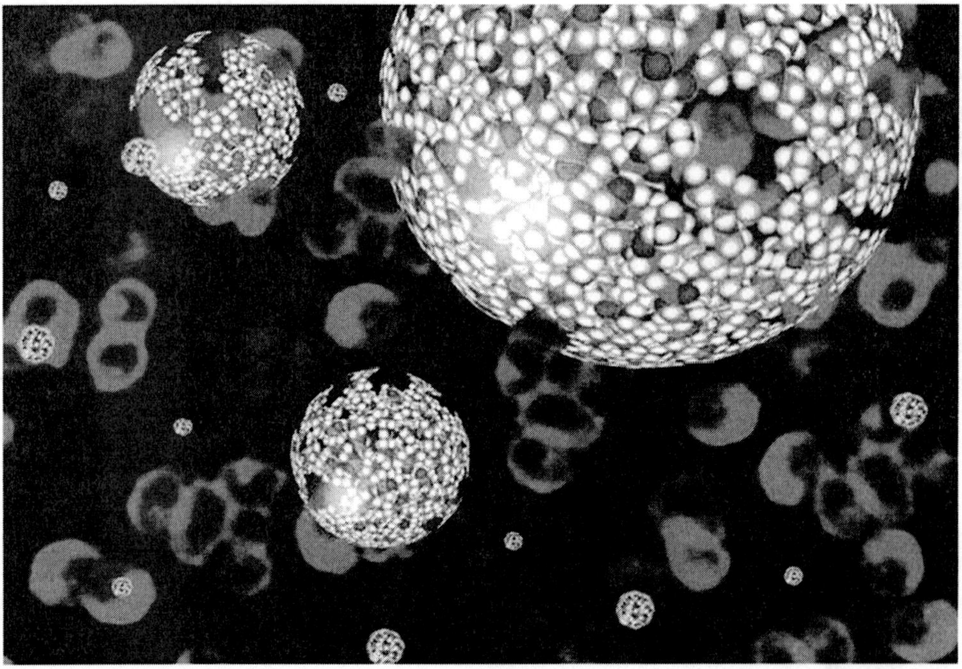

Figure: ***Biomedical Nanotechnology Research Thrust Area***

In the early 1960s, Alvin Weinberg, head of the Oak Ridge National Laboratory, was a strong advocate of nuclear power, and argued that it would provide "cheap energy." He was optimistic, but did his sums. First, he assumed that nuclear-power plants could be built a little more cheaply than coal-fired power plants of the same size. Then he assumed that the cost of fuel, waste disposal, operations, and maintenance for nuclear plants would be not much more than the cost of operations and maintenance alone for coal plants. Then he assumed that they might last for more than thirty years. Finally, he assumed that they would be publicly operated, tax free at low interest (which merely moves costs elsewhere) and that after thirty years the cost of the equipment would be written off (which is an accounting fiction). With all of that, he derived a power cost that "might be" as low as one half the cost of the cheapest coal-fired plant he mentions. He was clearly an optimist, but he didn't come close to arguing for power too cheap to meter.

STRONG AND LIGHTWEIGHT STRUCTURES

At the opposite extreme from molecular electronics-complex and at first worth billions of dollars per gram-are structural materials: worth only dollars per kilogram in most applications, but much simpler in structure. Once molecular manufacturing becomes inexpensive, structural materials will be important products.

These materials play a central role in almost everything around us, from cars and aircraft to furniture and houses. All of these objects get their size, shape, and strength from a structural skeleton of some sort. This makes structural materials a natural place to begin in understanding how nanotechnology can improve products.

Cars today are mostly made of steel, aircraft of aluminum, and buildings and furniture largely of steel and wood. These materials have a certain "strength-to-weight ratio" (more properly, a strength-to-density ratio). To make cars stronger, they'd have to be heavier; to make them lighter, they'd have to be weaker. Clever design can change this relationship a little, but to change it a lot requires a change of materials.

Making something heavy is easy: just leave a hollow space, then fill it with water, sand, or lead shot. Making something light and strong is harder, but often important. Automakers try to make cars lightweight, aircraft manufacturers try harder, and with spacecraft manufacturers it is an obsession. Reducing mass saves materials and energy.

The strongest materials in use today are mostly made of carbon. Kevlar, used in racing sails and bulletproof vests, is made of carbon-rich molecular fibers. Expensive graphite composites, used in tennis rackets and jet aircraft, are made using pure-carbon fibers. Perfect fibers of carbon-both graphite and diamond-would be even better, but can't be made with today's technology. Once molecular manufacturing gets rolling, though, such materials will be commonplace and inexpensive.

What will these materials be like? To picture them, a good place to start is wood. The structure of wood can vary from extremely light and porous, like balsa wood, to denser structures like oak. Wood is made by molecular machinery in plants from carbon-rich polymers, mostly cellulose. Molecular manufacturing will be able to make materials like these, but with a strength-to-weight ratio about a hundred times that of mediocre steel, and tens of times better than the best steel. Instead of being made of cellulose, these materials will be made of carbon in forms like diamond. Diamond is emphasized here not because it is shiny and expensive, but because it is strong and potentially cheap. Diamond is just carbon with properly arranged atoms. Companies are already learning to make it from natural gas at low pressure. Molecular manufacturing will be able to make complex objects of the stuff, built lighter than balsa wood but stronger than steel.

Products made of such materials could be startling by our present standards. Objects could be made that are identical in size and shape to those we make today, but simultaneously stronger and 90 percent lighter. This is something to keep in mind next time you're lugging a heavy object around. (If something needs weight to hold it in place, it would be more convenient to add this ballast when the thing is in its proper location than to build in the extra weight permanently.) Better structural materials will make aircraft lighter, stronger, and more efficient, but will have the greatest effect on spacecraft. Today, spacecraft can barely reach orbit with both a safety margin and a cargo. To get there at all, they have to drop off parts like boosters and tanks along the way, shedding weight.

Fast Development

In some areas of high technology-spaceflight has been a notorious example-it takes years, even decades, to try a new idea. This makes progress slow to a crawl. In other areas-software has been a shining example-new ideas can be tested in minutes or hours. Since the Space Shuttle design was frozen, personal computer software has come into existence and gone through several generations of commercial development, each with many cycles of building and testing.

Fast, Inexpensive Testing

Even in the days of the first operational molecular manipulators, experimentation is likely to be reasonably fast. Individual chemical steps can take seconds or less. Complex molecular objects could be built in a matter of hours. This will let new ideas be put into practice almost as fast as they can be designed.

Later assemblers will be even faster. At a millionth of a second per step, they will approach the speed of computers. And, as nanotechnology matures, experimenters will have more and more molecular instruments available to help them find out whether their devices work or not. Fast construction and fast testing will encourage fast progress.

At this point, the cost of materials and equipment for experiments will be trivial. No one today can afford to build Moon rockets on a hobby budget, but they can afford to build software, and many useful programs have been the result. There is no economic reason why nanomachines couldn't eventually be built with a hobby-size budget, though there are reasons-to be discussed in later chapters-for wanting to place limits on what can be built.

Prior Simplicity

Finally, established technologies are always pushing up against some limit; the easy opportunities have generally been exploited. In many fields, the limits are those of the properties of the materials used and the cost and precision of manufacturing. This is true for computers, for spacecraft, for cars, blenders, and shoes. For software, the limits are those of computer capacity and of sheer complexity (which is to say, of human intelligence). After molecular manufacturing develops certain basic abilities, a whole set of limits will fall, and a whole range of developments will become possible. Limits set by materials properties, and by the cost and precision of manufacturing, will be pushed way back. Competition, easy opportunities, and fast, low-cost experimentation should combine to yield an explosion of new products.

This does not mean immediately, and it does not apply to all imaginable nanotechnologies. Some technologies are imaginable and clearly feasible, yet dauntingly complex. Still, the above considerations suggest that a wide range of advances could happen at a brisk pace. The main bottleneck might seem to be a shortage of knowledgeable designers-hardly anyone knows both chemistry and mechanical design-but improving computer simulations will help. These simulations will let engineers tinker with molecular-machinery designs, absorbing knowledge of chemical rules without learning chemistry in the usual sense.

Climbing Complexity

Making familiar products from improved materials will increase their safety, performance, and usefulness. It will also present the simplest engineering task. A greater change, though, will result from unfamiliar products made possible by new manufacturing methods. In talking about unfamiliar products, a hard-to-answer question arises: What will people want?

Products are typically made because their recipients want them. In our discussions here, if we describe something that people won't want, then it probably won't get built, and if it does get built, it will soon disappear. (The exceptions-fraud, coercion, persistent mistakes-are important, but in other contexts.) To anchor our discussion, it makes sense to look not at totally new products, but instead at new features for old products, or new ways to provide old services. This approach won't cover more than a fraction of what is possible, but will start from something sensible and provide a springboard for the imagination. As usual, we are describing possibilities, not making predictions. The possibilities focused on here arise from more complex applications of molecular manufacturing-nanotechnological products that contain nanomachines when they are finished. Earlier, we discussed strong materials. Now, we discuss some smart materials.

The goal of making materials and objects smart isn't new: researchers are already struggling to build structures that can sense internal and environmental conditions and adapt themselves appropriately. There is even a Journal of Intelligent Material Systems and Structures. By using materials that can adapt their shapes, sometimes hooked up to sensors and computers, engineers are starting to make objects they call "smart." These are the early ancestors of the smart materials that molecular manufacturing will make possible.

Today, we are used to having machines with a few visible moving parts. In cars, the wheels go around, the windshield wipers go back and forth, the antenna may go up and down, the seat belts, mirrors, and steering wheel may be motor-driven. Electric motors are fairly small, fairly inexpensive, and fairly reliable, so they are fairly common. The result is machines that are fairly smart and flexible, in a clumsy, expensive way.

In the Desert Rose scenario, we saw "tents" being assembled from trillions of submicroscopically small parts, including motors, computers, fibers, and struts. To the naked eye, materials made from these parts could seem as smooth and uniform as a piece of plastic, or as richly textured as wood or cloth-it is all a matter of the arrangement and appearance of the submicroscopic parts. These motors and other parts cost less than a trillionth of a dollar apiece. They can be quite reliable, and good design can make systems work smoothly even if 10 percent of a trillion motors burn out. Likewise for motor-controlling computers and the rest. The resulting machines can be very smart and flexible, compared to those of today, and inexpensive, too. When materials can be full of motors and controllers, whole chunks of material can be made flexible and controllable. The applications should be broad.

Surfaces surround us, and human-made surfaces-walls, roofs, and pavement-cover huge areas that matter to people. How can smart materials make a difference here?

The revolution in technology has come and gone, and you want to repaint your walls. Breathing toxic solvents and polluting water by washing brushes have passed into history, because paint has been replaced with smarter stuff. The mid-twentieth century had seen considerable progress in paints, especially the development of liquids that weren't quite liquid-they would spread with a brush, but didn't (stupidly) run and drip under their own weight. This was an improvement, but the new material, "paperpaint," is even more cooperative.

Paperpaint comes in a box with a special trowel and pen. The paperpaint itself is a dry block that feels a lot like a block of wood. Following the instructions, you use the pen to draw a line around the edge of the area you want to paint, putting an X in the middle to show where you want the paint to go on; the line is made of nontoxic disappearing ink, so you can slop it around without staining anything. Using the trowel, you slice off a hunk of paperpaint-which is easy, because it parts like soft butter to the trowel, even though it behaves like a solid to everything else. Very high IQ stuff, that.

Next, you press the hunk against the X and start smoothing it out with the trowel. Each stroke spreads a wide swath of paperpaint, much wider than the trowel, but always staying within the inked line. A few swipes spreads it precisely to the edges, whereupon it smooths out into a uniform layer. Why doesn't it just spread itself? Experience showed that customers didn't mind the effort of making a few swipes and preferred the added control. The paperpaint consists of a huge number of nanomachines with little wheels for rolling over one another and little sticky pads for clinging to surfaces. Each has a simple, stupid computer on board. Each can signal its neigbours.

The whole mass of them clings together like an ordinary solid, but they can slip and slide in a controlled way when signaled. When you smooth the trowel over them, this contact tells them to get moving and spread out. When they hit the line, this tells them to stop. If they don't hit a line, they go a few handbreadths, then stop anyway until you trowel them again. When they encounter a line on all sides, word gets around, and they jostle around to form a smooth, uniform layer. Any that get scraped off are just so much loose dust, but they stick together quite well. This paint-stuff doesn't get anything wet, doesn't stain, and clings to surfaces just tightly enough to keep it from peeling off accidentally. If some experimentally minded child starts digging with a stick, makes a tear, and peels some off, it can be smoothed back again and will rejoin as good as new. The child may eat a piece, but careful regulation and testing has ensured that this is no worse than eating plain paper, and safer than eating a colorful Sunday newspaper page.

Many refinements are possible. Swipes and pats of the trowel could make areas thicken or thin, or bridge small holes (no more Spackling!). With sufficiently smart paperpaint, and some way to indicate what it should do, you can have your choice of textures. Any good design will be washable, and a better design would shed dirt automatically using microscopic brushes. Removal, of course, is easy: either you rip and peel (no scraping needed), or find that trowel, set the dial on the handle to "strip," and poke the surface a few times. Either way, you end up with a lump ready to pitch into the recycling bin and the same old wall you started with, bared to sight again.

Looking further at the human environment we find a lot of cloth and related materials, such as carpeting and shoes. The textile industry was at the cutting edge of the first industrial revolution, and the next industrial revolution will have its effects on textiles.

With nanotechnology, even the finest textile fibers could have sensors, computers, and motors in their core at little extra cost. Fabrics could include sensors able to detect light, heat, pressure, moisture, stress, and wear, networks of simple computers to integrate this data, and motors and other nanomechanisms to respond to it. Ordinary, everyday things like fabric and padding could be made responsive to a person's needs-changing shape, color, texture, fit, and so forth-with the weather and a person's posture or situation. This process could be slow, or it could be fast enough to respond to a gesture. One result would be genuine one-size-fits-all clothing (give or take child sizes), perfectly tailored off the rack, warm in winter, cool and dry in summer; in short, nanotechnology could provide what advertisers have only promised. Even bogus advertising gives a clue to human desires.

Throughout history, the human race has pursued the quest for comfortable shoes. With fully adjustable materials, the seemingly impossible goal of having shoes that both look good and feel good should finally be achieved. Shoes could keep your feet dry, and warm except in the Arctic, cool except in the tropics, and as comfortable as they can be with a person stepping on them.

Adaptive structures will be useful in furniture. Today, we have furniture that adapts to the human body, but it does so in an awkward and incomplete manner. It adapts because people grab cushions and move them around. Or a chair adapts because it is a hinged contraption that grudgingly bends and extends in a few places to suit a small range of preferred positions. Occasionally, one sees furniture that allegedly gives a massage, but in fact only vibrates.

These limitations are consequences of the expense, bulkiness, clumsiness, and unreliability of such things as moving parts, motors, sensors, and computers today. With molecular manufacturing, it will be easy to make furniture from smart materials that can adapt to an individual human body, and to a person's changing position, to consistently give comfortable support. Smart cushions could also do a better job of responding to hints in the form of pats, tugs, and punches. As for massage-a piece of furniture, no matter how advanced, is not the same as a masseuse. Still, a typical massage setting on a smart chair would not mean today's "vibrate medium vigorously," but something closer to "five minutes of shiatsu."

This tour through of the potential of smart matter has shown how we could get walls that look and sound as we wish, clothing, shoes, and furniture of greater comfort, and clean solar power. As one might expect, this just scratches the surface.

If you care to think of further applications, here are some ground rules: Components made by molecular manufacturing can be many tens of times stronger than steel, but materials made by plugging many components together will be weaker. For these, strengths in the range of cotton candy to steel seem achievable. The components will be sensitive to heat, and at high temperatures they will break down or burn. Many materials will be able to survive the temperature of boiling water, but only specialized designs would be oven-safe.

Color, texture, and (usually) sound should be controllable. Surfaces can be smooth and tightly sealed (this takes some cleverness). Motions can be fairly fast. Power has to come from somewhere; good sources include electricity, stored chemical energy, and light. If nanomachines or smart materials are dunked in liquids, chemical energy can come from dissolved molecules; if they are in the open, energy can come from light; if they are sitting in one place, they can be plugged into a socket; if they are moving around in the dark, they can run on batteries for a while, then run down and quit. Within these limits, much can be accomplished.

"Smart" is a relative term. Unless you want to assume that people learn a lot more about intelligence and programming, it is best to assume that these materials will follow simple rules, like those followed by parts of drawings on computer screens. In these drawings, a picture of a rectangle can be commanded to sprout handles at its corners; pulling a handle stretches or shrinks the rectangle without distorting its right-angle corners. An object made of smart matter could do likewise in the real world: a box could be stretched to a different size, then made rigid again; a door in a smart-material wall could have its position unlocked, its frame moved a pace to the left, and then be returned to normal use.

There seems little reason to make bits of smart matter independent, self-replicating, or toxic. With care, smart matter should be safer than what it replaces because it will be better controlled. Spray paint gets all over things and contains noxious solvents; the paperpaint described above doesn't. This will be a characteristic difference, if we exercise our usual vigilance to encourage the production of things that are safe and environmentally sound.

It may be fun to discuss wondrous new products, but they won't make much difference in the world if they are too expensive. Besides, many people today don't have decent food, clothes, and a roof over their heads, to say nothing of fancy "nanostuff."

Costs matter. There is more to life than material goods, but without material goods life is miserable and narrow. If goods are expensive, people strive for them; if goods are abundant, people can turn their attention elsewhere. Some of us like to think that we are above a concern for material goods, but this seems more common in the wealthy countries. Lowering manufacturing costs is a mundane concern, but so are feeding people, housing them, and building sewage systems to keep them from dying of cholera and hepatitis. For all these reasons, finding ways to bring down production costs is a worthy goal. For the poor, for the environment, and for the freeing of human potential, costs matter deeply. Let's take a closer look at the costs of molecular manufacturing.

Inflation produces the illusion that costs rise, when the real story is that the value of money is falling. In the short term, real costs usually don't change very quickly, and this can produce the illusion that costs are stable facts of nature, like the law of gravity or the laws of thermodynamics. In the real world, though, most costs have been falling by a crucial measure: the amount of human labor needed to make things. People can afford more and more, because their labor, supplemented by machines, can produce more and more. This change is dramatic measured on a scale of centuries, and equally dramatic across the gulf between Third World and developed countries. The rise from Third World to First World standards of living has raised income (dropped the cost of labor time) by more than a factor of ten. What can molecular manufacturing do?

Larger cost reductions have happened, most dramatically in computers. The cost of a computer of a given ability has fallen by roughly a factor of 10 every seven years since the 1940s. In total, this is a factor of a million. If automotive technologies had done likewise, a luxury car would now cost less than one cent. (Personal computer systems still cost hundreds of dollars both because they are far more powerful than the giant machines of the 1940s and because the cost of buying any useful computer system includes much more than just the cost of a bare computer chip.)

NANOTECHNOLOGY: THE NEXT LEAP

While the physicist-led team at U. Brob was finishing its work on the AFM-based molecular manipulator, a chemist-led team at the University of Lilliput was working furiously. They saw the U. Brob desktop machine as too large and its expected products as too expensive. Even back in the 1980s, David Biegelsen of the Xerox Palo Alto Research Center had noted, "The main drawback I see to using a hybrid protoassembler [AFM-based molecular manipulator] is that it would take a long time to build just one unit. Building requires a series of atom-by-atom construction steps. It would be better to build in parallel from the very beginning, making many trillions of these molecules all at the same time. I think there is tremendous power in parallel assembly. Maybe another field, chemistry or biology, offers a better way to do it." The chemists at U. Lill aimed to develop that better way, building first simple and then more and more complex molecular machines. The eventual result was a primitive molecular assembler able to build molecular objects by the trillions.

Chemist's Tools

How did the chemists achieve this? During the years when the U. Brob team was developing the molecular manipulator, researchers working in protein science and synthetic chemistry had made better and better systems of molecular building blocks. Chemists were well prepared for doing this: by the late 1980s, it had become possible to build stable structures the size of medium-sized protein molecules, and work had begun to focus on making these molecules perform useful work by binding and modifying other molecules. Chemists learned to use these sophisticated catalysts-early molecular devices-to make their own work easier by helping in the manufacture of still more large molecules.

Another traditional chemist's tool was software for doing computer-aided design. The early software designed by Jay Ponder and Frederic Richards of Yale University ultimately led to semi-automatic tools for designing molecules of a particular shape and function. Chemists then could easily design molecules that would self-assemble into larger structures, several tens of nanometers across.

MOLECULAR CONSTRUCTION MACHINES

These advances in software and chemical synthesis let the U. Lill team tackle the task of building a primitive version of a molecular assembler. Although they couldn't build anything as complex as a nanocomputer or as stiff as diamond, they didn't need to. Their design used sliding molecular rods to position a molecular gripper much like the gripper used at U. Brob, again using the surrounding liquid to control which tool the gripper held. Instead of an AFM's electronic controls, they used the surrounding liquid to control the position of the rods as well. In a neutral solution, the rods would withdraw; in an acid solution, they would extend. How far they moved depended on what other molecules were around to lodge in special pockets and block the motion.

Their primitive assemblers built much the same sorts of products that the U. Brob molecular manipulator did; the tools were similar, and speed and accuracy were about the same. Yet there was one dramatic advantage: About 1,000,000,000,000,000,000,000 U. Lill assemblers could fit in the space occupied by one U. Brob manipulator, and it was easy to produce a mere 1,000,000,000,000,000 times as much product at the same cost. With the first, primitive assemblers, construction was slow because each step required new liquid baths and several seconds of soaking and waiting, and a typical product might take thousands of steps. Nonetheless, the U. Lill team made a lot of money licensing their technology to researchers trying to commercialize products they had first researched with the U. Brob machine. After starting an independent company (Nanofabricators, Inc.), they poured their research efforts into building better machines.

Within a few years, they had assemblers with multiple grippers, each loaded with a different kind of tool; flashes of colored light would flip molecules from state to state (they copied these molecules from the pigments of the retina of the eye); flipping molecules would change tools and change rod positions.

Soaking and waiting become a thing of the past, and soon they were pouring out parts that, when mixed with liquid and added to dishes with special blank chips would build up the dense memory layers that made possible the Pocket Library.

That was when things started moving fast. The semiconductor industry went the way of the vacuum tube industry. Money and talent poured into the new technology.

Molecular CAD tools got better, assemblers made it easy to build what was designed, and fast production and testing made molecular engineering as easy as playing with software. Assemblers got better, faster, and cheaper. Researchers used assemblers to build nanocomputers, and nanocomputers to control better, faster assemblers. Using tools to build better tools is an ancient story. Within a decade, almost anything could be made by molecular manufacturing, and was.

PROPERTIES OF NANOMATERIALS

In tandem with surface-area effects, quantum effects can begin to dominate the properties of matter as size is reduced to the nanoscale. These can affect the optical, electrical and magnetic behaviour of materials, particularly as the structure or particle size approaches the smaller end of the nanoscale. Materials that exploit these effects include quantum dots, and quantum well lasers for optoelectronics.

For other materials such as crystalline solids, as the size of their structural components decreases, there is much greater interface area within the material; this can greatly affect both mechanical and electrical properties. For example, most metals are made up of small crystalline grains; the boundaries between the grain slow down or arrest the propagation of defects when the material is stressed, thus giving it strength.

If these grains can be made very small, or even nanoscale in size, the interface area within the material greatly increases, which enhances its strength. For example, nanocrystalline nickel is as strong as hardened steel.

Understanding surfaces and interfaces is a key challenge for those working on nanomaterials, and one where new imaging and analysis instruments are vital.

NANOMATERIAL SCIENCE

Nanomaterials are not simply another step in the miniaturization of materials. They often require very different production approaches. There are several processes to create nanomaterials, classified as 'top-down' and 'bottom-up'. Although many nanomaterials are currently at the laboratory stage of manufacture, a few of them are being commercialised. Below we outline some examples of nanomaterials and the range of nanoscience that is aimed at understanding their properties. As will be seen, the behaviour of some nanomaterials is well understood, whereas others present greater challenges.

NANOSCALE IN TWO DIMENSIONS

Two dimensional nanomaterials such as tubes and wires have generated considerable interest among the scientific community in recent years. In particular, their novel electrical and mechanical properties are the subject of intense research.

Carbon Nanotubes

Carbon nanotubes (CNTs) were first observed by Sumio Iijima in 1991. CNTs are extended tubes of rolled graphene sheets.

Models of different singlewall nanotubes

There are two types of CNT: single-walled (one tube) or multi-walled (several concentric tubes). Both of these are typically a few nanometres in diameter and several micrometres to centimetres long. CNTs have assumed an important role in the context of nanomaterials, because of their novel chemical and physical properties. They are mechanically very strong (their Young's modulus is over 1 terapascal, making CNTs as stiff as diamond), flexible (about their axis), and can conduct electricity extremely well (the helicity of the graphene sheet determines whether the CNT is a semiconductor or metallic). All of these remarkable properties give CNTs a range of potential applications: for example, in reinforced composites, sensors, nanoelectronics and display devices. CNTs are now available commercially in limited quantities. They can be grown by several techniques. However, the selective and uniform production of CNTs with specific dimensions and physical properties is yet to be achieved. The potential similarity in size and shape between CNTs and asbestos fibres has led to concerns about their safety.

Inorganic Nanotubes

Inorganic nanotubes and inorganic fullerene-like materials based on layered compounds such as molybdenum disulphide were discovered shortly after CNTs. They have excellent tribological (lubricating) properties, resistance to shockwave impact, catalytic reactivity, and high capacity for hydrogen and lithium storage, which suggest a range of promising applications. Oxide-based nanotubes (such as titanium dioxide) are being explored for their applications in catalysis, photo-catalysis and energy storage.

Nanowires

Nanowires are ultrafine wires or linear arrays of dots, formed by self-assembly. They can be made from a wide range of materials. Semiconductor nanowires made of silicon, gallium nitride and indium phosphide have demonstrated remarkable optical, electronic and magnetic characteristics (for example, silica nanowires can bend light around very tight corners). Nanowires have potential applications in high-density data storage, either as magnetic read heads or as patterned storage media, and electronic and opto-electronic nanodevices, for metallic interconnects of quantum devices and nanodevices.

The preparation of these nanowires relies on sophisticated growth techniques, which include selfassembly processes, where atoms arrange themselves naturally on stepped surfaces, chemical vapour deposition (CVD) onto patterned substrates, electroplating or molecular beam epitaxy (MBE). The 'molecular beams' are typically from thermally evaporated elemental sources.

Biopolymers

The variability and site recognition of biopolymers, such as DNA molecules, offer a wide range of opportunities for the self-organization of wire nanostructures into much more complex patterns. The DNA backbones may then, for example, be coated in metal. They also offer opportunities to link nano- and biotechnology in, for example, biocompatible sensors and small, simple motors.

Such self-assembly of organic backbone nanostructures is often controlled by weak interactions, such as hydrogen bonds, hydrophobic, or van der Waals interactions (generally in aqueous environments) and hence requires quite different synthesis strategies to CNTs, for example. The combination of one-dimensional nanostructures consisting of biopolymers and inorganic compounds opens up a number of scientific and technological opportunities.

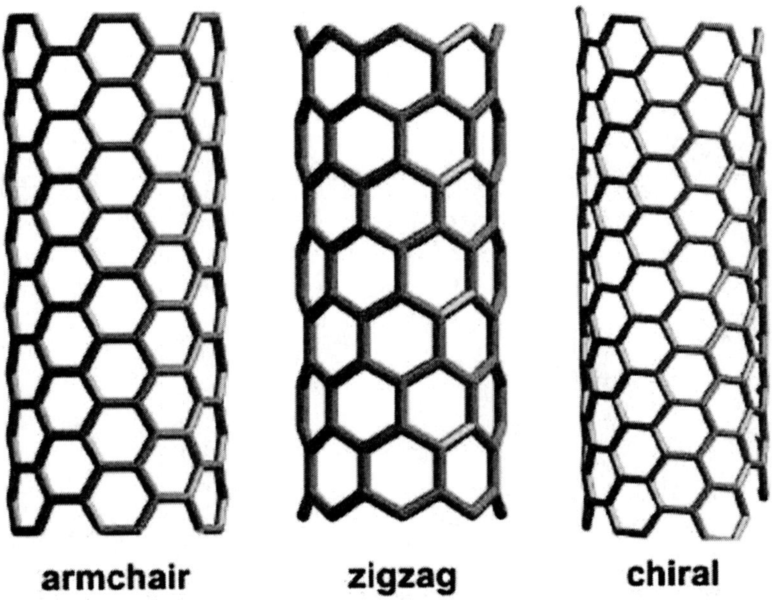

*Figure: **Single-walled nanotubes***

NANOSCALE IN THREE DIMENSIONS

Nanoparticles

Nanoparticles are often defined as particles of less than 100nm in diameter. We classify nanoparticles to be particles less than 100nm in diameter that exhibit new or enhanced size-dependent properties compared with larger particles of the same material. Nanoparticles exist widely in the natural world: for example as the products of photochemical and volcanic activity, and created by plants and algae. They have also been created for thousands of years as products of combustion and food cooking, and more recently from vehicle exhausts. Deliberately manufactured nanoparticles, such as metal oxides, are by comparison in the minority. Nanoparticles are of interest because of the new properties (such as chemical reactivity and optical behaviour) that they exhibit compared with larger particles of the same materials. For example, titanium dioxide and zinc oxide become transparent at the nanoscale, however are able to absorb and reflect UV light, and have found application in sunscreens. Nanoparticles have a range of potential applications: in the short-term in new cosmetics, textiles and paints; in the longer term, in methods of targeted drug delivery where they could be to used deliver drugs to a specific site in the body.

Nanoparticles can also be arranged into layers on surfaces, providing a large surface area and hence enhanced activity, relevant to a range of potential applications such as catalysts. Manufactured nanoparticles are typically not products in their own right, but generally serve as raw materials, ingredients or additives in existing products. Nanoparticles are currently in a small number of consumer products such as cosmetics and their enhanced or novel properties may have implications for their toxicity. For most applications, nanoparticles will be fixed (for example, attached to a surface or within in a composite) although in others they will be free or suspended in fluid. Whether they are fixed or free will have a significant affect on their potential health, safety and environmental impacts.

Fullerenes (carbon 60)

In the mid-1980s a new class of carbon material was discovered called carbon 60 (C60).Harry Kroto and Richard Smalley, the experimental chemists who discovered C60 named it "buckminsterfullerene", in recognition of the architect Buckminster Fuller, who was well-known for building geodesic domes, and the term fullerenes was then given to any closed carbon cage. C60 are spherical molecules about 1nm in diameter, comprising 60 carbon atoms arranged as 20 hexagons and 12 pentagons: the configuration of a football. In 1990, a technique to produce larger quantities of C60 was developed by resistively heating graphite rods in a helium atmosphere. Several applications are envisaged for fullerenes, such as miniature 'ball bearings' to lubricate surfaces, drug delivery vehicles and in electronic circuits.

Dendrimers

Dendrimers are spherical polymeric molecules, formed through a nanoscale hierarchical self-assembly process. There are many types of dendrimer; the smallest is several nanometres in size. Dendrimers are used in conventional applications such as coatings and inks, but they also have a range of interesting properties which could lead to useful applications. For example, dendrimers can act as nanoscale carrier molecules and as such could be used in drug delivery. Environmental clean-up could be assisted by dendrimers as they can trap metal ions, which could then be filtered out of water with ultra-filtration techniques.

Quantum Dots

Nanoparticles of semiconductors (quantum dots) were theorized in the 1970s and initially created in the early 1980s.

If semiconductor particles are made small enough, quantum effects come into play, which limit the energies at which electrons and holes (the absence of an electron) can exist in the particles. As energy is related to wavelength (or colour), this means that the optical properties of the particle can be finely tuned depending on its size.

Thus, particles can be made to emit or absorb specific wavelengths (colours) of light, merely by controlling their size. Recently, quantum dots have found applications in composites, solar cells (Gratzel cells) and fluorescent biological labels (for example to trace a biological molecule) which use both the small particle size and tuneable energy levels.

Recent advances in chemistry have resulted in the preparation of monolayer-protected, high-quality, monodispersed, crystalline quantum dots as small as 2nm in diameter, which can be conveniently treated and processed as a typical chemical reagent.

APPLICATIONS OF NANOMATERIALS

Below we list some key current and potential shortand long-term applications of nanomaterials. Most current applications represent evolutionary developments of existing technologies: for example, the reduction in size of electronics devices.

Current Applications

Nanosized titanium dioxide and zinc oxide are currently used in some sunscreens, as they absorb and reflect ultraviolet (UV) rays and yet are transparent to visible light and so are more appealing to the consumer. Nanosized iron oxide is present in some lipsticks as a pigment but it is our understanding that it is not used by the European cosmetics sector. The use of nanoparticles in cosmetics has raised a number of concerns about consumer safety.

Composites

An important use of nanoparticles and nanotubes is in composites, materials that combine one or more separate components and which are designed to exhibit overall the best properties of each component. This multifunctionality applies not only to mechanical properties, but extends to optical, electrical and magnetic ones. Currently, carbon fibres and bundles of multi-walled CNTs are used in polymers to control or enhance conductivity, with applications such as antistatic packaging.

The use of individual CNTs in composites is a potential long-term application. A particular type of nanocomposite is where nanoparticles act as fillers in a matrix; for example, carbon black used as a filler to reinforce car tyres. However, particles of carbon black can range from tens to hundreds of nanometres in size, so not all carbon black falls within our definition of nanoparticles.

Clays

Clays containing naturally occurring nanoparticles have long been important as construction materials and are undergoing continuous improvement. Clay particle based composites – containing plastics and nano-sized flakes of clay – are also finding applications such as use in car bumpers.

Coatings and Surfaces

Coatings with thickness controlled at the nano- or atomic scale have been in routine production for some time, for example in molecular beam epitaxy or metal oxide chemical vapor depositionfor optoelectonic devices, or in catalytically active and chemically functionalized surfaces.

Recently developed applications include the self-cleaning window, which is coated in highly activated titanium dioxide, engineered to be highly hydrophobic (water repellent) and antibacterial, and coatings based on nanoparticulate oxides that catalytically destroy chemical agents.

Wear and scratch-resistant hard coatings are significantly improved by nanoscale intermediate layers (or multilayers) between the hard outer layer and the substrate material.

The intermediate layers give good bonding and graded matching of elastic and thermal properties, thus improving adhesion.

A range of enhanced textiles, such as breathable, waterproof and stainresistant fabrics, have been enabled by the improved control of porosity at the nanoscale and surface roughness in a variety of polymers and inorganics.

Tougher and Harder Cutting Tools

Cutting tools made of nanocrystalline materials, such as tungsten carbide, tantalum carbide and titanium carbide, are more wear and erosion-resistant, and last longer than their conventional (large-grained) counterparts. They are finding applications in the drills used to bore holes in circuit boards.

Short-term Applications

Paints

Incorporating nanoparticles in paints could improve their performance, for example by making them lighter and giving them different properties. Thinner paint coatings ('lightweighting'), used for example on aircraft, would reduce their weight, which could be beneficial to the environment. However, the whole life cycle of the aircraft needs to be considered before overall benefits can be claimed. It may also be possible to substantially reduce solvent content of paints, with resulting environmental benefits. New types of foulingresistant marine paint could be developed and are urgently needed as alternatives to tributyl tin (TBT), now that the ecological impacts of TBT have been recognised.

Anti-fouling surface treatment is also valuable in process applications such as heat exchange, where it could lead to energy savings. If they can be produced at sufficiently low cost, fouling-resistant coatings could be used in routine duties such as piping for domestic and industrial water systems. It remains speculation whether very effective anti-fouling coatings could reduce the use of biocides, including chlorine. Other novel, and more long-term, applications for nanoparticles might lie in paints that change colour in response to change in temperature or chemical environment, or paints that have reduced infra-red absorptivity and so reduce heat loss.

Concerns about the health and environmental impacts of nanoparticles may require the need for the durability and abrasion behaviour of nano-engineered paints and coatings to be addressed, so that abrasion products take the form of coarse or microscopic agglomerates rather than individual nanoparticles.

Remediation

The potential of nanoparticles to react with pollutants in soil and groundwater and transform them into harmless compounds is being researched. In one pilot study the large surface area and high surface reactivity of iron nanoparticles were exploited to transform chlorinated hydrocarbons (some of which are believed to be carcinogens) into less harmful end products in groundwater. It is also hoped that they could be used to transform heavy metals such as lead and mercury from bioavailable forms into insoluble forms. Serious concerns have been raised over the uncontrolled release of nanoparticles into the environment.

Fuel Cells

Engineered surfaces are essential in fuel cells, where the external surface properties and the pore structure affect performance. The hydrogen used as the immediate fuel in fuel cells may be generated from hydrocarbons by catalytic reforming, usually in a reactor module associated directly with the fuel cell. The potential use of nano-engineered membranes to intensify catalytic processes could enable higher-efficiency, small-scale fuel cells. These could act as distributed sources of electrical power. It may eventually be possible to produce hydrogen locally from sources other than hydrocarbons, which are the feedstocks of current attention.

Displays

The huge market for large area, high brightness, flat-panel displays, as used in television screens and computer monitors, is driving the development of some nanomaterials.

Nanocrystalline zinc selenide, zinc sulphide, cadmium sulphide and lead telluride synthesized by sol–gel techniques (a process for making ceramic and glass materials, involving the transition from a liquid 'sol' phase to a solid 'gel' phase) are candidates for the next generation of light-emitting phosphors. CNTs are being investigated for low voltage field-emission displays; their strength, sharpness, conductivity and inertness make them potentially very efficient and long-lasting emitters.

Batteries

With the growth in portable electronic equipment (mobile phones, navigation devices, laptop computers, remote sensors), there is great demand for lightweight, high-energy density batteries. Nanocrystalline materials synthesized by sol–gel techniques are candidates for separator plates in batteries because of their foam-like (aerogel) structure, which can hold considerably more energy than conventional ones. Nickel–metal hydride batteries made of nanocrystalline nickel and metal hydrides are envisioned to require less frequent recharging and to last longer because of their large grain boundary (surface) area.

Fuel Additives

Research is underway into the addition of nanoparticulate ceria (cerium oxide) to diesel fuel to improve fuel economy by reducing the degradation of fuel consumption over time.

Catalysts

In general, nanoparticles have a high surface area, and hence provide higher catalytic activity. Nanotechnologies are enabling changes in the degree of control in the production of nanoparticles, and the support structure on which they reside. It is possible to synthesise metal nanoparticles in solution in the presence of a surfactant to form highly ordered monodisperse films of the catalyst nanoparticles on a surface. This allows more uniformity in the size and chemical structure of the catalyst, which in turn leads to greater catalytic activity and the production of fewer byproducts.

It may also be possible to engineer specific or selective activity. These more active and durable catalysts could find early application in cleaning up waste streams. This will be particularly beneficial if it reduces the demand for platinum-group metals, whose use in standard catalytic units is starting to emerge as a problem, given the limited availability of these metals.

Longer-term Applications

CNTs have exceptional mechanical properties, particularly high tensile strength and light weight. An obvious area of application would be in nanotubereinforced composites, with performance beyond current carbon-fibre composites. One current limit to the introduction of CNTs in composites is the problem of structuring the tangle of nanotubes in a well-ordered manner so that use can be made of their strength. Another challenge is generating strong bonding between CNTs and the matrix, to give good overall composite performance and retention during wear or erosion of composites.

The surfaces of CNTs are smooth and relatively unreactive, and so tend to slip through the matrix when it is stressed. One approach that is being explored to prevent this slippage is the attachment of chemical side-groups to CNTs, effectively to form 'anchors'. Another limiting factor is the cost of production of CNTs. However, the potential benefits of such light, high strength material in numerous applications for transportation are such that significant further research is likely.

Lubricants

Nanospheres of inorganic materials could be used as lubricants, in essence by acting as nanosized 'ball bearings'. The controlled shape is claimed to make them more durable than conventional solid lubricants and wear additives. Whether the increased financial and resource cost of producing them is offset by the longer service life of lubricants and parts remains to be investigated. It is also claimed that these nanoparticles reduce friction between metal surfaces, particularly at high normal loads. If so, they should find their first applications in high-performance engines and drivers; this could include the energy sector as well as transport. There is a further claim that this type of lubricant is effective even if the metal surfaces are not highly smooth. Again, the benefits of reduced cost and resource input for machining must be compared against production of nanolubricants. In all these applications, the particles would be dispersed in a conventional liquid lubricant; design of the lubricant system must therefore include measures to contain and manage waste.

Magnetic Materials

It has been shown that magnets made of nanocrystalline yttrium–samarium–cobalt grains possess unusual magnetic properties due to their extremely large grain interface area (high coercivity can be obtained because magnetization flips cannot easily propagate past the grain boundaries).

This could lead to applications in motors, analytical instruments like magnetic resonance imaging (MRI), used widely in hospitals, and microsensors. Overall magnetisation, however, is currently limited by the ability to align the grains' direction of magnetisation. Nanoscale-fabricated magnetic materials also have applications in data storage.

Devices such as computer hard disks depend on the ability to magnetize small areas of a spinning disk to record information.

If the area required to record one piece of information can be shrunk in the nanoscale (and can be written and read reliably), the storage capacity of the disk can be improved dramatically. In the future, the devices on computer chips which currently operate using flows of electrons could use the magnetic properties of these electrons, called spin, with numerous advantages. Recent advances in novel magnetic materials and their nanofabrication are encouraging in this respect.

Medical Implants

Current medical implants, such as orthopaedic implants and heart valves, are made of titanium and stainless steel alloys, primarily because they are biocompatible. Unfortunately, in some cases these metal alloys may wear out within the lifetime of the patient. Nanocrystalline zirconium oxide (zirconia) is hard, wearresistant, bio-corrosion resistant and bio-compatible. It therefore presents an attractive alternative material for implants. It and other nanoceramics can also be made as strong, light aerogels by sol–gel techniques. Nanocrystalline silicon carbide is a candidate material for artificial heart valves primarily because of its low weight, high strength and inertness.

Machinable Ceramics

Ceramics are hard, brittle and difficult to machine. However, with a reduction in grain size to the nanoscale, ceramic ductility can be increased. Zirconia, normally a hard, brittle ceramic, has even been rendered superplastic (for example, able to be deformed up to 300% of its original length). Nanocrystalline ceramics, such as silicon nitride and silicon carbide, have been used in such automotive applications as high-strength springs, ball bearings and valve lifters, because they can be easily formed and machined, as well as exhibiting excellent chemical and high-temperature properties. They are also used as components in high-temperature furnaces. Nanocrystalline ceramics can be pressed into complex net shapes and sintered at significantly lower temperatures than conventional ceramics.

Water Purification

Nano-engineered membranes could potentially lead to more energy-efficient water purification processes, notably in desalination by reverse osmosis. Again, these applications would represent incremental improvements in technologies that are already available. They would use fixed nanoparticles, and are therefore distinct from applications that propose to use free nanoparticles.

Military Battle Suits

Enhanced nanomaterials form the basis of a state-of- the-art 'battle suit' that is being developed by the Institute of Soldier Nanotechnologies at MIT. A short-term development is likely to be energy-absorbing materials that will withstand blast waves; longer-term are those that incorporate sensors to detect or respond to chemical and biological weapons (for example, responsive nanopores that 'close' upon detection of a biological agent). There is speculation that developments could include materials which monitor physiology while a soldier is still on the battlefield, and uniforms with potential medical applications, such as splints for broken bones.

NANOTECHNOLOGY- FROM LITTLE THINGS, BIG CONUNDRUMS GROW

Whether you fear it, welcome it, don't understand it or think it an elaborate scientific conspiracy, "molecular nanotechnology" (MNT), or "molecular manufacturing," will have a profound and instantaneous effect on all aspects of industry and society. According to the Center for Responsible Nanotechnology (CRN), the new technology will hit us with all the destructive force of a speeding freight train, and "without adequate preparation and study, the effects could be dangerously disruptive." Yet more science hype? Perhaps. But in 11 essays, published by Nanotechnology Perceptions, a peer-reviewed academic journal of the Collegium Basilea in Basel, Switzerland, a number of highly regarded scientists cite past mistakes involving the introduction of a new technology without adequate public consultation and discussion. They also imagine a number of future nanotech scenarios that both fascinate and frighten. In one of the essays, Singularities and Nightmares, David Brin offers some positive words on the potential benefits of nanotechnology, such as genome mapping while-you-wait. Perhaps, he says, this could be achieved conveniently by: "compressing a complete biochemical laboratory the size of a house down to something that fits cheaply on your desktop. A MolecuMac, if you will." But on the flip side, Brin makes your hair stand on end when he discusses the possibility of MolecuMac "hackers", whose ranks may include terrorists, despots or even angry, misunderstood, spotty-faced youths hell-bent on making your life a misery. "What are we going to do when kids all over the world can analyze and synthesize any organic compound, at will? What happens when this kind of [hacking] 'creativity' moves to the very stuff of life itself?" asks Brin.

Brin draws attention to influential people in society, such as Francis Fukuyama, author of The End of History, whose opinions are polar-opposite when it comes down to the issue of how much control and regulation there should be on technological innovation.

Renunciation is the key word for this group, who believe that in order to save humanity from itself we must show restraint and curtail our creative juices for the sake of a better world. Brin says that authors such as Margaret Attwood and Michael Crichton often express: "outright loathing for the overweening arrogance of hubristic technological innovators who just cannot leave nature well enough alone." And what about former Sun Computers exec Bill Joy, who claims that the notion of an open society is dead? Brin says that Joy uses the Unabomber as reason enough for believing that: "our sole hope for survival may be to renounce, squelch, or relinquish several classes of technological progress." Phew. Brin is supportive of the social critics who shout when they see potential danger along the road, but he doesn't take the argument for containment as far as they do.

Most of the CRN fraternity seems to support the happy medium Brin refers to as "reciprocal accountability"; which advocates openness and accountability for technological development. Firebrand futurologist Ray Kurzweil sums it up when he says: "As the pace of technological advancement rapidly accelerates, it becomes increasingly important to promote knowledgeable and insightful discussion of both promise and peril." This is probably true of any new technology, but especially true in regard to nanotechnology, claims Mike Treder, the Executive Director of the CRN. Which is why, in August 2005, CRN formed a special task force comprised of experts from multiple disciplines whose job it is to: "develop comprehensive recommendations for the safe and responsible use of nanotechnology". But is a collaborative network of big-brained individuals really enough to affect how the future of molecular manufacturing will pan out? The fact is, most other attempts at creating an effective policy to govern new technology have failed or never seen the light of day, because nobody can really predict what applications will stem from new scientific discoveries. And it is this freedom to create without prophetic insight that gets us all into so much trouble. So it stands to reason that there will be some who think humanity is not yet responsible enough to have that freedom. If history is anything to go by, perhaps the neo-luddites are correct and we must make technological sacrifices to save ourselves, as the halfway policy of reciprocal accountability just won't cut it.

It would be hard to imagine, for example, President Truman saying to the A-bomb team, "ok, boys, wrap it up, I'm losing sleep over this technology." And what about all those private institutes, whose doors are more often than not closed to public or government scrutiny? Just what will be in your face cream or toothpaste in the years to come? It's hard to imagine just how a nanotech policy would work in advance of any future applications. And who's to say that in trying to prevent the development of an undesirable application, a policy panel wouldn't also stifle positive innovation and progress? The industrial age seems to have a lot to answer for in regard to climate change these days, but would any of us have prevented the development of the internal combustion engine? Should we have turned our backs on electricity, because Harold P. Brown invented the electric chair? And would we have sacrificed radioactivity at the expense of some of the more advanced medical technologies? It's an unfortunate fact that very often the applications stemming from scientific discoveries are Janus faced. And unfortunately, no god has endowed anyone from CRN with second sight.

If nanotechnology proves to live up to all the hype, and personal nanofactories become the norm, it will be the most versatile and pervasive technology ever conceived by humanity. As such, it is unlikely that any meaningful guidelines could be imposed upon research and development for any future molecular manufacturing applications. While the CRN's Task Force is a welcome initiative, regulating potential applications once molecular manufacturing is in full swing will become analogous to bailing out a sinking boat with a sieve.

RNA Used as Nano-Scaffolding

Microscopic scaffolding to house the tiny components of nanotech devices could be built from ribonucleic acid (RNA), the same substance that transports messages around a cell's nucleus, says a research group at Purdue University. By encouraging RNA molecules to self-assemble into 3-D shapes resembling spirals, triangles, rods and hairpins, the group has found what could be a method of constructing lattices on which to build complex microscopic machines. From such RNA blocks, the group has already constructed arrays that are several micrometers in diameter.

"Our work shows that we can control the construction of three-dimensional arrays made from RNA blocks of different shapes and sizes," said Peixuan Guo, from Purdue's School of Veterinary Medicine. "With further research,

RNA could form the superstructures for tomorrow's nanomachines." The research appears in the journal Nano Letters. Nanotechnologists, like those in Guo's group, hope to build microscopic devices with sizes that are best measured in nanometers- or billionths of a meter. Because nature routinely creates nano-sized structures for living things, many researchers are turning to biology for their inspiration and construction tools. "Biology builds beautiful nanoscale structures, and we'd like to borrow some of them for nanotechnology," Guo said. "The trouble is, when we're working with such tiny blocks, we are short of tiny steam shovels to push them around. So we need to design and construct materials that can assemble themselves."

Organisms are built in large part of three main types of building blocks: proteins, DNA and RNA. Of the three, perhaps least investigated and understood is RNA, a molecular cousin to the DNA that stores genetic blueprints within our cells' nuclei. RNA typically receives less attention than other substances from many nanotechnologists, but Guo said the molecule has distinct advantages. "RNA combines the advantages of both DNA and proteins and puts them at the nanotechnologist's disposal," Guo said. "It forms versatile structures that are also easy to produce, manipulate and engineer."

Since his discovery of a novel RNA that plays a vital role in a microscopic "motor" used by the bacterial virus phi29, Guo has continued to study the structure of this RNA molecule for years. It formed the "pistons" of a tiny motor his lab created several years ago, and members of the team collaborated previously to build dimers and trimers- molecules formed from two and three RNA strands, respectively. "By designing sets of matching RNA molecules, we can program RNA building blocks to bind to each other in precisely defined ways," he said. "We can get them to form the nano-shapes we want."

From the small shapes that RNA can form- hoops, triangles and so forth- larger, more elaborate structures can in turn be constructed, such as rods gathered into spindly, many-pronged bundles. These structures could theoretically form the scaffolding on which other components, such as nano-sized transistors, wires or sensors, could be mounted.

"Because these RNA structures can be engineered to put themselves together, they could be useful to industrial and medical specialists, who will appreciate their ease of engineering and handling," said co-researcher Dieter Moll. "Self-assembly means cost-effective."

Moll, while bullish on RNA's prospects, cautioned that there was more work to be done before nanoscale models could be built at will. "One of our main concerns right now is that, over time, RNA tends to degrade biologically," he said. "We are already working on ways to make it more resistant to degradation so that it can form long-lasting structures."

"We have not built actual scaffolds yet, just 3-D arrays," he said. "But we have built them from engineered biological molecules, and that could help us bridge the gap between the living and the nonliving world. If nanotech devices can eventually be built from both organic and inorganic materials, it would ease their use in both medical and industrial settings, which could multiply their usefulness considerably."

NANOELECTRONICS NANOTECHNOLOGY

Experts everywhere say that nanotechnology will lead to smarter, faster, and more efficient electronics, but they also admit that researchers face some challenging obstacles along the information superhighway.

Particle contamination is one of the thorniest problems facing the electronics and computing industries. Even the tiniest particle can destroy a computer chip and cause other damage. Several University researchers are searching for ways to detect particle contaminants and prevent their formation.

A team of researchers headed by mechanical engineering professor Peter McMurry is studying the formation and growth of contaminants in semiconductor processing equipment. Smaller contaminated particles deposit more quickly than larger ones and can destroy a number of memory chips and processors. The team's goal is to design new processes or tools for producing contaminant-free devices.

"It's important to understand how to prevent particle deposition, and this will require an understanding of how such particles are formed and transported," McMurry says.

His team has developed a device called the particle beam mass spectrometer, the only available instrument for measuring nanoparticles in low-pressure semiconductor processing equipment.

Electrical engineering professor Steve Campbell is among the researchers who are looking for ways to create electronic devices with greater memory. His studies are aimed at detecting and preventing the formation of particle contaminants in chemical reactors. Although research in the area is still relatively new, some early prototypes have been made, says Campbell.

Campbell's Microtechnology Laboratory team has developed a machine to detect and measure particles. "Now we're trying to demonstrate that isolating these particles and using them to our advantage is feasible," he says, adding that his group is seeking funding to continue its research.

"The importance of smaller and faster semiconductor devices is evident from the trends we've all seen [toward] faster, more powerful, and less costly computers," says McMurry. "Work that is being done to fabricate devices at the nanoscale will ensure that this trend will continue for some time to come."

The same is true in recording media. Associate Professor Randall Victora and Professor Jack Judy of the electrical and computer engineering department received grants two years ago from the National Storage Industry Consortium and Seagate Corporation to develop superlattice magnetic recording media. The focus of their research is magnetic recording technology for hard drives, which are vulnerable to thermal fluctuations. These fluctuations can affect the recording process and destroy vital information stored on a disk in the drive.

"We're preventing that from happening," Victora says. "These superlattices cause the disk to have a stronger resistance to these thermal fluctuations."

Once developed, the technology would be used primarily by the computing industry. Companies such as Quantum, IBM, and Seagate already have expressed interest in the group's research, Victora says.

Other University projects also have implications for the electronics industry, including the study of adhesion in computer chip manufacturing, molecular electronics, and the search for an appropriate nonconductor to replace copper, which is limited in its ability to transport information between resistors.

DNA SCAFFOLDING

The next frontier for information processing may lie at the interface of nanoelectronics and biotechnology. An interdisciplinary team led by electrical and computer engineering professor Richard Kiehl is exploring the use of DNA as a programmable scaffolding for the self-assembly of nanoscale electronic components. As a model for fabricating and designing semiconductor devices and circuits, DNA offers two key advantages: size scale and programmability.

Most industry experts believe that within the next 10 to 15 years the ability to scale down conventional technologies will reach its limit. At that point, the operating principles of conventional devices-and the techniques used to fabricate them-will break down. The basic elements of the DNA molecule are at just the right scale, says Kiehl.

Self-assembly uses bio-recognition, a natural process in which one molecule is attracted to and binds with another to form small structures. In the case of DNA, the attraction can be programmed so that the molecules will spontaneously assemble in solution to achieve a desired result.

"It's possible to synthesize small versions of DNA molecules in the laboratory and program in whatever code you want," says Kiehl. "And because the two strands of DNA have complementary codes that match up, you can design one strand of DNA in a certain way so it will match another strand and assemble a nanoscale structure this way."

The matched segments form a scaffolding on which nanoparticles are affixed at highly selective attachment points. It's an approach that offers the programmability and precision needed for assembling electronic circuitry on the nanoscale. "We have to make a real paradigm shift," Kiehl says. "Not only do we have to keep improving performance, but we also must look at the kinds of devices we can make at those scales and how we want to use them to process information." To that end, the researchers are turning to the human brain for inspiration. They envision devices whose electrical characteristics resemble those of neuron-like electrical waveforms in the brain. Like certain regions of the brain, the devices would process information based on pattern recognition rather than on individual bits of information. It's a more sophisticated level of information processing than can be achieved using conventional computers. Kiehl predicts there will be a wide range of applications for this technology, including signal processing, communications systems, and computer systems. "The higher end of this [work] will be things that computers can't do very well today because the operations they use are too restrictive. One is the ability to recognize a pattern, such as identifying a letter as being an 'A' or a 'B', or being able to identify a face. "It won't be just making things faster and faster in the conventional way," he says. "It will really be opening up new ways to process information in machines."

THE GENE GUN

David Pui has big dreams, all generated by an ultrasmall technology, and he's not alone. Leaders from industry, academia, and government, including President Bill Clinton and University president Mark Yudof, agree that nanotechnology could someday revolutionize science, medicine, and industry.

Pui, a Distinguished McKnight University Professor in mechanical engineering, is director of the Particle Technology Laboratory-a hub of nanotechnology research at the University-and a staunch advocate of the emerging field. Pui, who chaired the Minnesota Nanotechnology Summit last March, says that the University has already established itself as a leader in promoting and supporting interdisciplinary nanotechnology research.

Pui should know. He's been involved in one of the University's most promising nanotechnology achievements to date, the development of a "gene gun"-a newly patented electrospraying apparatus for continuous gene transfection. The gun shows great potential as a delivery method for gene therapy.

Nowhere are the hopes for nanotechnology riding higher than among medical researchers. The gene gun project is the result of years of research by Pui and Assistant Professor Da-Ren Chen, also a member of the mechanical engineering faculty. Christine Wendt, an assistant professor in the Medical School who has worked with Pui's team, says she can readily foresee ways in which the gun could have medical applications. The goal of gene therapy is to treat cells with genes that will help the cells produce specific proteins to fight disease. But in order for gene therapy to work, a number of things must happen. The gene first must get to the cell, and then the cell must accept it. Finally, the gene must arrive at the nucleus for its effects to take hold.

Traditional methods for this process, called gene transfection, include uptaking plasmid DNA, infecting the cell with viruses that contain the desired gene, fusing cells together, and microinjecting cells. All these methods treat a large number of cells simultaneously, but they tend to have a low efficiency rate. In treating lung cancer, for example, doctors have been frustrated by their inability to successfully deliver genes to diseased cells.

"The traditional ways of treating many diseases have limitations, but gene delivery offers new and exciting possibilities," Wendt says. "When you use chemicals to insert genes into cells, there are efficiency and price issues. Shocking the cell with electricity can cause serious damage, and it's a method that's limited to cells contained in petri dishes. The important thing is to optimize gene transfer carefully, which this gene gun can do. It has the potential to be applied directly to the treatment of human disease."

The physical transfer of genes is not a new concept; however, the technology behind the University's gene transfector is novel and offers distinct advantages over its precursors.

In the late 1980s, horticultural scientist John Sanford and his colleagues at Cornell University designed a gene gun that has been widely used, most commonly in the genetic transformation of plants. To deliver genes to the targeted cells, the Sanford gun uses a bullet-like projectile coated with gold particles that act as gene carriers.

The gun itself consists of a highly pressurized upper chamber and a low-pressure lower chamber with a diaphragm in the middle. When the diaphragm is punctured, pressure from the upper chamber emits a shock wave that hits the projectile, discharging it forward until the projectile hits a porous screen. The force of the impact launches the gold particles through the screen downstream toward the targeted cells. The particles hit the cells at a velocity high enough to puncture the cell membrane, penetrate the cells, and release the genes, which are diffused into the nucleus.

According to the University researchers, the Sanford gun has a number of disadvantages. The individual gold particles that coat the projectile are harmless, but they have a tendency to stick together and crush the cell, an effect called "pit damage."

Sometimes researchers have to shoot the same cell sample repeatedly to obtain the desired result. After each shot, the diaphragm and projectile must be reloaded and the process repeated. If there are thousands of cells to shoot, the preparation of the gene-coated carriers becomes a grueling, time-consuming process.

Although it's similar in concept to the Sanford device, the University's gun uses a very different technology, a patented process called continuous gene transfection. A gene suspension-with or without gold particles-is loaded into a capillary via a syringe. The suspension is sprayed out of the capillary at a controlled pace under a high electric field, which is created by applying voltage. The spray itself consists of highly charged and dispersed gene particles or gene-coated particles. Particles with a similar charge repel each other at a velocity high enough to penetrate the cell membrane and release the genes into the cells.

"It's a simple process," says Chen, who is credited with designing the electrospray process. "But it's just how you engineer it to get it to do what you want it to do." The University researchers also had to be careful not to infringe upon the Sanford gun's patent. Specifically, they had to avoid a technology that used external forces to discharge the gene carriers into the cells, as the Sanford gun does. Pui and his team had to devise a different technique.

Furthermore, they had to design the gun with a specific application in mind. That issue was resolved after Wendt met Pui and Chen, who discussed their ideas for the gene gun project with her. Wendt, who was working on another project involving the effects of small particles, says she knew instantly that the gene gun had great potential for application in her field.

She worked closely with Pui and Chen in testing the gun. In one experiment, the researchers injected enhanced green fluorescent proteins (EGFP) from jellyfish into different types of cells, including African green monkey cells and human lung-cancer cells. The EGFP proteins glow when exposed to ultraviolet light, so the researchers could see whether or not the genes had entered the cells.

Experiments also proved that the gun's design and technology offered other advantages over existing methods of gene transfection. The charged particles repel each other with such force that there's no risk of pit damage. The gun's capillary-and-syringe setup permits continuous injection of the genes. In fact, Pui says, the process could run continuously for days before the gene suspension had to be reloaded. It's a method that could be adapted to a system of mass production, in which the target cells or samples are placed on a conveyor belt and transported to the spraying assembly.

"We can shoot until it's sufficient," Chen says. "[With the Sanford gun], they have to shoot sometimes five times, sometimes up to 30 times, and that can really be a headache. Plus it takes a while to prepare the gene formula on the projectile. When they have to change, it is a lengthy process." Because their gene gun worked successfully on different types of cells, Pui and his team believe that it can be used on an even broader range of cell types. The experiments also proved that the gold particles weren't essential to the technology and that the process could be conducted on a mass scale.

The years of design and experimentation have paid off. On July 25, Pui and Chen received U.S. patent No. 6,093,557 for an "electrospraying apparatus and method for introducing material into cells." It's one of three patents Pui and Chen have received for their nanoparticle research over the past two years. The other patents, issued last fall, are for devices used to charge the particles. Another patent request is in the works, but the researchers are still cautious about discussing it.

"We've demonstrated that the devices work," Pui says. "They have been proven, and because of that, outside interest has been high." He says one local company has licensed the nanoparticle charger patents and another large Minnesota company has expressed interest in the gene gun-related devices.

Although test results so far have been very successful, the gun is still in the earliest stages of development. Pui and his team have identified several major goals for the next phase of their research, which will subject the device to more difficult challenges. Besides improving the gun's design and dosage controls, they intend to use it on more-difficult cell types and experiment with different genes.

The team also wants to expand the range of the gun's applications by making it portable. Theoretically, doctors treating lung cancer could insert a portable gun into the patient's lung and spray the diseased cells. Once inside the lung, the gun would release the gene, which the patient then could inhale.

"We need to continue to work on this gene gun project," Pui says. "Furthermore, we need to continue to work with other departments, like we did with Christine (Wendt). This work crosses many fields."

It's work that requires both the visionary's capacity for big dreams and the scientist's meticulous, ceaseless scrutiny. But above all, it's the dreaming that compels researchers like Pui, Chen, and Wendt to probe deeper into the tiny world of nanotechnology. And their research has already given them a taste of what lies ahead.

As Chen says, "Now we have to explore all of the possibilities."

BUILDING WITH MOLECULAR GRIPPERS AND TOOLS

To build something with the U. Brob team's AFM-based molecular manipulator system, you use it as follows: First, choose a surface to build on and place it under the tip in a pool of liquid. Then dunk the AFM tip into the liquid, bringing it down to the surface, and back it off a little. Construction can now begin as soon as a tool is loaded into the gripper.

Tubes and pumps can flow different liquids over the surface and past the gripper, carrying different tool molecules. If you want to do something with a tool of Type A, you wash in the proper liquid. Once it is in the gripper, you can use the AFM mechanism to move it around and put it where you want it. Move it up to the surface at a convenient spot, wait a few seconds, and it reacts, forming a bond and leaving a molecular fragment attached to the spot you chose. To add a different fragment, you can use a tool of Type B: you back up the tip, flow in a fresh liquid carrying the new tools, and in a moment a tool of the new type is bound in place and ready to apply, either on or alongside the first spot. Step by step, you build up a precise molecular structure.

Each step takes only seconds. Molecular tools pop into the gripper in a fraction of a second, and used tools pop off at the same rate. Once the tip has positioned a molecule, it reacts quickly, about a million times faster than unwanted reactions at other sites. In this way, the molecular manipulator gives good control of where reactions will occur (though it is not as reliable as an advanced assembler would be). It is fairly fast by a chemist's standards—per cycle—but still a million times slower than an advanced assembler. It can perform a variety of steps, but isn't as flexible and capable as an advanced assembler. In short, it is hardly the last word in nanotechnology, yet is a great advance over what has gone before.

Products

With its ability to accelerate desired reactions by a factor of a million or so, the U. Brob team's molecular manipulator can perform 10,000 to 100,000 steps with good reliability. Back in the 1980s, chemists making protein molecules struggled to perform just one hundred steps. The U. Brob research team (and its many imitators) can now build structures that are stronger and easier to design than proteins: not floppy, folded chains, but rugged objects held together by a sturdy network of bonds. Though not as strong and dense as diamond, these structures are like bits of a tough engineering plastic. A specially adapted computer-aided design system makes it easy to design molecular objects made from these materials.

Yet the AFM-based molecular manipulator has one grave disadvantage: It does chemistry one molecule at a time, and it ties up a machine as expensive as a car for hours or days to produce that one large molecule. Some molecules, though, are valuable enough to be worth building even one at a time. These draw prompt attention.

A single molecule isn't much use as a dye, a drug, or a floor wax, but it can have substantial value if it provides useful information. The U. Brob team quickly publishes a pile of scientific papers based on experiments with single molecules: they build a molecule, probe it, report the results, and build another. Some of these results show chemists elsewhere in the multibillion-dollar chemical industry how to design new catalysts, molecules that can help make other molecules more cheaply, cleanly, and efficiently. This information is worth a lot.

Three new products of special interest are among the first to be made. The first—molecular electronics—begins with experiments conducted by a research group at a computer chip company. They use their molecular manipulator to build single molecules and probe them, gradually learning how to build the parts needed for molecular electronic computers. These new computers don't immediately become practical, because the costs are too high for making such large molecules with AFM-based technology. Yet some companies begin to produce simpler molecular electronic devices for use in sensors and specialized high-speed signal processing. A specialty industry is born and begins to expand.

The second product is a gene reader, a complex molecular device built on the surface of a chip. The biologists who built the reader combined proteins borrowed from cells with special-purpose molecular machines designed from scratch. The result was a molecular system that binds DNA molecules and pulls them past a read-head-like tape through a tape recorder. The device works as fast as some naturally occurring molecular machines that read DNA, with one key advantage: it outputs its data electronically. At that speed, a single device can read a human genome in about a year. Though still too expensive for a doctor's office, these readers are promptly in great demand from research laboratories. Another small industry is born.

The third product is far more important, in the long run: replacement tips for molecular manipulators, grippers, and tools that are better than the originals. With these new, more versatile devices, researchers are now building more ambitious products and tools.

3

The Nanomanufacturing

Nanomanufacturing is used to describe either the production of nano scaled materials, which can be powders or fluids, or describe the manufacturing of parts "button up" from nanoscaled materials or "top down" in smallest steps for high precison by several technologis like laser ablation, etching and others.

The term is widely used, e.g. by the European Technology Platform MINAM and the U.S. National Nanotechnology Initiative (NNI). NNI refer to a sub-effort of nanotechnology--one of its five "priority areas."

There is also a Nanomanufacturing Program at the U.S. National Science Foundation. Nanomanufacturing should not be confused with molecular manufacturing which refers specifically to the manufacture of complex, nanoscale structures by means of nonbiological mechanosynthesis (and subsequent assembly).

The NNI has defined nanotechnology very broadly, to include a wide range of tiny structures including those created by large and imprecise tools.

Nanomanufacturing is not defined in the NNI's recent report, "Instrumentation and Metrology for Nanotechnology." (For contrast, another "priority area," nanofabrication, is defined as "the ability to fabricate, by directed or self-assembly methods, functional structures or devices at the atomic or molecular level." p. 67) Nanomanufacturing appears to be the near-term industrial-scale manufacture of nanotechnology-based objects, with emphasis on low cost and reliability.

THE APPLICATIONS

The development of devices that are small, light, self-contained, use little energy and that will replace larger microelectronic equipment is one of the first goals of the anticipated nanotechnology revolution. The second phase will be marked by the introduction of materials not feasible at larger than nanotechnology levels. Given the nature of quantum variance, scientists theorize that single molecule sensors can be developed and that sophisticated memory storage and neural-like networks can be achieved with a very small number of molecules. Traditional engineering concepts undergo radical transformation at the atomic level. For example, nano-technology motors may drive gears, the cogs of which are composed of the atoms attached to a carbon ring. Nanomotors may themselves be driven by oscillating magnetic fields or high precision oscillating lasers.

Perhaps the greatest promise for nanotechnology lies in potential biotechnology advances. Potential nano-level manipulation of DNA offers the opportunity to radically expand the horizons of genomic medicine and immunology. Tissue-based biosensors may unobtrusively be able to monitor and regulate site-specific medicine delivery or regulate physiological processes. Nanosystems might serve as highly sensitive detectors of toxic substances or used by inspectors to detect traces of biological or chemical weapons. In electronics and computer science, scientists assert that nanotechnologies will be the next major advance in computing and information processing science. Microelectronic devices rely on recognition and flips in electron gating (e.g. where differential states are ultimately represented by a series of binary numbers ["0" or "1"] that depict voltage states). In contrast, future quantum processing will utilize the identity of quantum states as set forth by quantum numbers. In quantum cryptography systems with the ability to decipher encrypted information will rely on precise knowledge of manipulations used to achieve various atomic states. Nanoscale devices are constructed using a combination of fabrication steps. In the initial growth stage, layers of semiconductor materials are grown on a dimension limiting substrate. Layer composition can be altered to control electrical and/or optical characteristics. Techniques such as molecular beam epitaxy (MBE) and metallo-organic chemical vapor deposition (MOCVD) are capable of producing layers of a few atoms thickness. The developed pattern is then imposed on successive layers (the pattern transfer stage) to develop desired three dimensional structural characteristics.

NANOCOMPUTERS

While the first, primitive assemblers were controlled by changing what molecules are in the solution around the device, getting the speed and accuracy wanted for large-scale manufacturing takes real computation. Carl's setup uses a combination of special-purpose molecule processors and general-purpose assemblers, all controlled and orchestrated by nanocomputers.

Computers back in the 1990s used microelectronics. They worked by moving electrical charge back and forth through conducting paths-wires, in effect-using it to block and unblock the flow of charge in other paths. With nanotechnology, computers are built from molecular electronics. Like the computers of the 1990s, they use electronic signals to weave the patterns of digital logic. Being made of molecular components, though, they are built on a much smaller scale than 1990s computers, and work much faster and more efficiently. On the scale of our simulated molecular world, 1990s computer chips are like landscapes, while nanocomputers are like individual buildings. Carl's desktop PC contains over a trillion nanocomputers, enough to out-compute all the microelectronic computers of the twentieth century put together.

Back in the dark ages of the 1980s, an exploratory engineer proposed that nanocomputers could be mechanical, using sliding rods instead of moving electrons as shown in Figure 8. These molecular mechanical computers were much easier to design than molecular electronic computers would have been. They were a big help in getting some idea of what nanotechnology could do.

An electronic transistor lets current flow when a negative electric charge is applied and blocks current when a positive charge is applied. The mechanical "transistor" lets the horizontal rod move when the vertical rod is down, and blocks the horizontal rod when the vertical rod is up. Either device can be used to build logic gates and computers.

Even back then, it was pretty obvious that mechanical computers would be slower than electronic computers. Carl's molecular electronic PC would have been no great surprise, though nobody knew just how to design one. When nanotechnology actually arrived and people started competing to build the best possible computers, molecular electronics won the technology race. Still, mechanical nanocomputers could have done all the nanocomputing jobs at Desert Rose: ordinary, everyday molecular manufacturing just doesn't demand the last word in computer performance.

For Carl, the millions of nanocomputers in the milky waters of his building ponds are just extensions of machines on his desk, machines there to help him run his business and deliver products to his customers-or, in the case of the Red Cross emergency, to help provide time-critical emergency supplies. By reserving those three separate ponds, Carl can either build three different kinds of equipment for the Red Cross or use all the ponds to mass-produce the first thing on the Red Cross list: emergency shelters for ten thousand people. The software is ready, the plumbing is fine, the drums of building materials are all topped up, the Special Mix for this job is loaded: the build is ready to start. "Okay," Carl tells the computer, "build Red Cross tents." Computer talks to nanocomputers. In all three pools, nanocomputers talk to assemblers. The build begins.

ASSEMBLING The PRODUCTS

Some of the building done at Desert Rose Industries uses assemblers much like the ones we saw in the first hall of the plant tour, back in the simulated molecular world of the Silicon Valley Faire. As seen in simulation, they are big, slow, computer-controlled things moving molecular tools. With the right instructions and machinery to keep them supplied with molecular tools, these general-purpose assemblers can build almost anything. They're slow, though, and take a lot of energy to run. Some of the building uses special-purpose assembly systems in the molecule-processing style, like the systems in the basement we saw in the tour of a simulated molecular factory. The special-purpose systems are all moving belts and rollers, but no arms. This is faster and more efficient, but for quantity orders, cooling requirements limit the speed.

It's faster to use larger, prefabricated building blocks. Desert Rose uses these for most of their work, and especially for rush orders like the one Carl just set up. Their underground warehouse has room-sized bins containing upward of a thousand tons of the most popular building blocks, things like structural fibers. They're made at plants on the West Coast and shipped here by subway for ready use. Other kinds are made on site using the special-purpose assemblers. Carl's main room has several cabinet-sized boxes hooked up to the plumbing, each taking in raw materials, running them through this sort of specialized molecular machinery, and pumping out a milky syrup of product. One syrup contains motors, another one contains computers, and another is full of microscopic plug-in light sources. All go into tanks for later use.

Now they're being used. The mix for the Red Cross tent job is mostly structural fiber stronger than the old bulletproof-vest materials. Other building blocks also go in, including motors, computers, and dozens of little struts, angle brackets, and doohickies. The mix would look like someone had stirred together the parts from a dozen toy sets, if the parts were big enough to see. In fact, though, the largest parts would be no more than blurry dots, if you saw one under a normal optical microscope.

The mix also contains block-assemblers, floating free like everything else. These machines are big, about like an office building in our simulation view with the standard settings. Each has several jointed arms, a computer, and several plugs and sockets. These do the actual construction work.

To begin the build, pumps pour the mix into a manufacturing pond. The constant tumbling motions of microscopic things in liquids would be too disorganized for building anything so large as a tent, so the block-assemblers start grabbing their neigbours. Within moments, they have linked up to form a framework spread through the liquid. Now that they are plugged together, they divide up jobs, and get to work. Instructions pour in from Carl's desktop computer.

The block-assemblers use sticky grippers to pull specific kinds of building blocks out of the liquid. They use their arms to plug them together. For a permanent job, they would be using blocks that bond together chemically and permanently. For these temporary tents, though, the Red Cross design uses a set of standard blocks that are put together with amazingly ordinary fasteners: these blocks have snaps, plugs, and screws, though of course the parts are atomically perfect and the threads on the screws are single helical rows of atoms. The resulting joints weaken the tent's structure somewhat, but who cares? The basic materials are almost a hundred times stronger than steel, so there is strength to waste if it makes manufacturing more convenient.

Fiber segments snap together to make fabrics. Some segments contain motors and computers, linked by fibers that contain power and data cables. Struts snap together with more motors and computers to make the tent's main structures. Special surfaces are made of special building blocks. From the human perspective, each tent is a lightweight structure that contains most of the conveniences and comforts of an apartment: cooking facilities, a bathroom, beds, windows, air conditioning, specially modified to meet the environmental demands of the quake-stricken country. From a builder's perspective, especially from a nanomachine's point of view, the tent is just structure slapped together from a few hundred kinds of prefab parts.

In a matter of seconds, each block-assembler has put together a few thousand parts, and its section of the tent is done. In fact, the whole thing is done: many trillions of hands make light work. A crane swings out over the pond and starts plucking out tent packages as fresh mix flows in.

Maria's concern has drawn her back to the plant to see how the build is going. "It's coming along," Carl reassures her. "Look, the first batch of tents is out." In the warehouse, the first pallet is already stacked with five layers of dove-gray "suitcases": tents dried and packed for transport. Carl grabs a tent by the handle and lugs it out the door. He pushes a tab on the corner labeled "Open," and it takes over a minute to unfold to a structure a half-dozen paces on a side. The tent is big, and light enough to blow away if it didn't cling to the ground so tightly. Maria and Carl tour the tent, testing the appliances, checking the construction of furniture: everything is extremely lightweight compared to the bulk-manufactured goods of the 1990s, tough but almost hollow.

Like the other structures, the walls and floors are full of tiny motors and struts controlled by simple computers like the ones used in twentieth-century cars, televisions, and pinball machines. They can unfold and refold. They can also flex to produce sound like a high-quality speaker, or to absorb sound to silence outdoors racket. The whole three-room setup is small and efficient, looking like a cross between a boat cabin and a Japanese business hotel room. Outside, though, it is little more than a box. Maria shakes her head, knowing full well what architects can do these days when they try to make a building really fit its site. Oh well, she thinks, These won't be used for long.

Behind the Scenes and Afterward

Desert Rose Industries and other manufacturers can make almost anything quickly and at low cost. That includes the tunneling machines and other equipment that made the subway system they use for shipping. Digging a tunnel from coast to coast now costs less than digging a single block under New York City used to. It wasn't expensive to get a deep-transit terminal installed in their basement. Just as the tents aren't mere bundles of canvas, these subways aren't slow things full of screeching, jolting metal boxes. They're magnetically levitated to reach aircraft speeds-as experimental Japanese trains were in the late 1980s-making it easy for Carl and Maria to give their customers quick service. There's still a road leading to the plant, but nobody's driven a truck over it for years.

They only take in materials that they will eventually ship out in products, so there's nothing left over, and no wastes to dump. One corner of the plant is full of recycling equipment. There are always some obsolete parts to get rid of, or things that have been damaged and need to be reworked. These get broken down into simpler molecules and put back together again to make new parts.

The gunk in the manufacturing ponds is water mixed with particles much finer than silt. The particles-fasteners, computers, and the rest-stay in suspension because they are wrapped in molecular jackets that keep them there. This uses the same principle as detergent molecules, which coat particles of oily dirt to float them away.

Though it wouldn't be nutritious or appetizing, you could drink the tent mix and be no worse for it. To your body, the parts and their jackets, and even the nanomachines, would be like so many bits of grit and sawdust. (Grandma would have called it roughage.)

Carl and Maria get their power from solar cells in the road, which is the only reason they bothered having it paved. In back of their plant stands what looks like a fat smokestack. All it produces, though, is an updraft of clean, warm air. The darkly paved road, baking in the New Mexico sun, is cooler than you might expect: it soaks up solar energy and makes electricity, instead of just heat. Once the power is used, it turns back into heat, which has to go somewhere. So the heat rises from their cooling tower instead of the road, and the energy does useful work on the way.

Some products, like rocket engines, are made more slowly and in a single piece. This makes them stronger and more permanent. The tents, though, don't need to be superstrong and are just for temporary use. A few days after the tents go up, the earthquake victims start to move out into new housing (permanent, better-looking, and very earthquake resistant). The tents get folded and shipped off for recycling.

Recycling things built this way is simple and efficient: nanomachines just unscrew and unsnap the connectors and sort the parts into bins again. The shipments Desert Rose gets are mostly recycled to begin with. There's no special labeling for recycled materials, because the molecular parts are the same either way.

For convenience (and to keep the plant small), Carl and Maria get most of their parts prefabricated, even though they can make almost anything. They can even make more production equipment. In one of their manufacturing ponds, they can put together a new cabinet full of special-purpose assemblers. They do this when they want to make a new type of part in-house. Like parts, the part-assemblers are made by special-purpose assemblers. Carl can even makc big vats in medium-size vats, unfolding them like tents.

If Desert Rose Industries needed to double capacity, Carl and Maria could do it in just a few days. They did this once for a special order of stadium sections. Maria got Carl to recycle the new building before its shadow hurt their cactus garden.

In the Desert Rose Industries scenario, manufacturing has become cheap, fast, clean, and efficient. Using fast, precise machines to handle matter in molecular pieces makes it easy for nanotechnology to be fast, clean, and efficient. But for it to be cheap, the manufacturing equipment has to be cheap. The Desert Rose scenario shows how this can work. Molecular-manufacturing equipment can be used to make all the parts needed to build more molecular manufacturing equipment. It can even build the machines needed to put the parts together. This resembles an idea developed by NASA for a self-expanding manufacturing complex on the Moon, but made faster and simpler using molecular machines and parts.

In the early days of nanotechnology, there won't be as many different kinds of machines as there are at Desert Rose. One way to build a lot of molecular manufacturing equipment in a reasonable time would be to make a machine that can be used to make a copy of itself, starting with special but simple chemicals. A machine able to do this is called a "replicator." With a replicator and a pot full of the right fuel and raw materials, you could start with one machine, then have two, four, eight, and so on.

This doubling process soon makes enough machines to be useful. The replicators-each including a computer to control it and a general-purpose assembler to build things-could then be used to make something else, like the tons of specialized machines needed to set up a Desert Rose manufacturing plant. At that point, the replicators could be discarded in favour of those more efficient machines.

Replicators are worth a closer look, though, because they show how quickly molecular manufacturing systems can be used to build more manufacturing equipment. If we were in one of our standard simulation views, the submicroscopic device at the top of the picture would be like a huge tank, three stories tall when lying on its side. Most of its interior is taken up by a tape memory system that tells how to move the arm to build all the parts of the replicator, except the tape itself. The tape gets made by a special tape-copying machine. At the right-hand end of the replicator are pores for bringing in fuel and raw-material molecules, and machinery for processing them. In the middle are computer-controlled arms, like the ones we saw on the plant trip. These do most of the actual construction.

The steps in the cycle-using a copy to block the tube, beginning a fresh copy, then releasing the old one-illustrate one way for a machine to build a copy of itself while floating in a liquid, yet doing all its construction work inside, in vacuum. (It's easier to design for vacuum, and this is exploratory-engineering work, so easier design is better design.) Calculations suggest that the whole construction cycle can be completed in less than a quarter hour, since the replicator contains about a billion atoms, and each arm can handle about a million atoms per second. At that rate, one device can double and double again to make trillions in about ten hours.

Each replicator just sits in a chemical bath, soaking up what it needs and making more replicators. Eventually, either the special chemicals run out or other chemicals are added to signal them to do something else. At that point, they can be reprogrammed to produce anything else you please, so long as it can be extruded from the front. The products can be long, and can unfold or be pieced together to make larger objects, so the size of these initial replicators-smaller than a bacterium-would be only a temporary limitation.

From the molecular manipulators and primitive assemblers described in the last chapter, the most likely path to nanotechnology leads to assemblers with more and more general capabilities. Still, efficiency favours special-purpose machines, and the Desert Rose scenario didn't make much use of general assemblers. Why bother making general-purpose assemblers in the first place?

To see the answer, turn the question around and ask, Why not build such a tool? There is nothing outstandingly difficult about a general assembler, as molecular machinery goes. It will just be a device with good, flexible positional control and a system to feed it a variety of molecular tools. This is a useful, basic capability. General-purpose assemblers could always be replaced by a lot of specialized devices, but to build those specialized devices in the first place, it makes sense to come up with a more flexible, general-purpose system that can just be reprogrammed.

So, general purpose machines are likely to find use in making short production runs of more specialized devices. Ralph Merkle, a computers and security expert at Xerox Palo Alto Research Center, sees this as paralleling the way manufacturing works today: "General purpose devices could do many tasks, but they'll do them inefficiently. For any given task, there will be one or a few best ways of doing it, and one or a few special purpose devices that are finely tuned to do that one task. Nails aren't made by a general-purpose machine shop, they're made by nail-making machines. Making nails with a general-purpose machine shop would be more expensive, more difficult, and more time-consuming. Likewise, in the future we won't see a proliferation of general-purpose self-replicating systems, we'll see specialization for almost every task."We've surveyed a lot of devices: assemblers of various flavors, nanocomputers, disassemblers, replicators, and others. What's important about these is not the exact distinctions between them, but the capabilities that they will give and the effects they will have on human lives. Again, we are suspending discussion of potential misapplications until later.

If we tease apart the implications of what we've seen in the Desert Rose scenario, we can analyze some of the key impacts of molecular manufacturing in industry, science, and medicine.

THE INDUSTRY AND TECHNOLOGY

At its base, nanotechnology is about molecular manufacturing, and manufacturing is the basis of much of today's industry. This is why Desert Rose made a good starting point for describing the possibilities of a nanotechnological world. From an industrial perspective, it makes sense to think of nanotechnology in terms of products and production.

New Products: Today, we handle matter crudely, but nanotechnology will bring thorough control of the structure of matter, the ability to build objects to atom-by-atom specifications. This means being able to make almost anything. By comparison, even today's range of products will feel very limited. Nanotechnology will make possible a huge range of new products, a range we can't envision today. Still, to get a feel for what is possible, we can look at some easily imagined applications.

Reliable Products: Today, products often fail, but for failures to occur-for a wing to fall off an airplane, or a bearing to wear out-a lot of atoms have to be out of place. In the future, we can do better. There are two basic reasons for this: better materials and better quality control, both achieved by molecular manufacturing. By using materials tens of times stronger than steel, as Desert Rose did, it will be easy to make things that are very strong, with a huge safety margin. By building things with atom-by-atom control, flaws can be made very rare and extremely small-nonexistent, by present standards.

With nanotechnology, we can design in big safety margins and then manufacture the design with near-perfection. The result will be products that are tough and reliable. (There will still be room for bad designs, and for people who wish to take risks in machines that balance on the edge of disaster.)

Intelligent Products: Today, we make most things from big chunks of metal, wood, plastic, and the like, or from tangles of fibers. Objects made with molecular manufacturing can contain trillions of microscopic motors and computers, forming parts that work together to do something useful. A climber's rope can be made of fibers that slide around and reweave to eliminate frayed spots. Tents can be made of parts that slide and lock to turn a package into a building. Walls and furniture can be made to repair themselves, instead of passively deteriorating. On a mundane level, this sort of flexibility will increase reliability and durability. Beyond this, it will make possible new products with abilities we never imagined we needed so badly. And beyond even this, it will open new possibilities for art.

Inexpensive Production: Today, production requires a lot of labor, either for making things or for building and maintaining machines that make things. Labor is expensive, and expensive machines make automation expensive, too. In the Desert Rose scenario, we got a glimpse of how molecular manufacturing can make production far less expensive than it is today. This is perhaps the most surprising conclusion about nanotechnology, so we'll take a closer look at it in the next chapter.

Clean Production: Today, our manufacturing processes handle matter sloppily, producing pollution. One step puts stuff where it shouldn't be; the next washes it off the product and into the water supply. Our transportation system worsens the problem as unreliable trucks and tankers spill noxious chemicals over the land and sea. Everything is expensive, so companies skimp on even the half-effective pollution controls that we know how to build.

Nanotechnology will mean greater control of matter, making it easy to avoid pollution. This means that a little public pressure will go a long way toward a cleaner environment. Likewise, it will make it easy to increase efficiency and reduce resource requirements. Products, like the Red Cross tents at Desert Rose, can be made of snap-together, easily recyclable parts. Sophisticated products could even be made from biodegradable materials. Nanotechnology will make it easy to attack the causes of pollution at their technological root.

Nanotechnology will have great applications in the field of industry, much as transistors had great applications in the field of vacuum tube electronics, and democracy had great applications in the field of monarchy. It will not so much advance twentieth-century industry as replace it-not all at once, but during a thin slice of historical time.

Chemistry: Today, chemists work with huge number of molecules and study them using clever, indirect techniques. Making a new molecule can be a major project, and studying it can be another. Molecular manufacturing will help chemists make what they want to study, and it will help them make the tools they need to study it. Nanoinstruments will be used to prod, measure, and modify molecules in a host of ways, studying their structures, behaviors, and interactions.

Materials: Today, materials scientists make new superconductors, semiconductors, and structural materials by mixing and crushing and baking and freezing, and so forth. They dream of far more structures than they can make, and they stumble across more things than they plan. With molecular manufacturing, materials science can be much more systematic and thorough. New ideas can be tested because new materials can be built according to plan (rather than playing around, groping for a recipe).

This need not rule out unexpected discoveries, since experiments-even blind searches-will go much faster. A few tons of raw materials would be enough to make a billion samples, each a cubic micron in size. In all of history so far, materials scientists have never tested so many materials. With nanoinstruments and nanocomputers, they could. One laboratory could then do more than all of today's materials scientists put together.

Biology: Today, biologists use a host of molecular devices borrowed from biology to study biology. Many of these can be viewed as molecular machines. Nanotechnology will greatly advance biology by providing better molecular devices, better nanoinstruments. Some cells have already been mapped in amazing molecular detail, but biology still has far to go. With nanoinstruments (including molecule-by-molecule disassemblers), biologists will at last be able to map cells completely and study their interactions in detail. It will become easy not only to find molecules in cells, but to learn what they do. This will help in understanding disease and the molecular requirements for health, enormously advancing medicine.

Computation: Today, computers range from a million to a billion times faster than an old desktop adding machine, and the results have been revolutionary for science. Every year, more questions can be answered by calculations based on known principles of physics. The advent of nanocomputers-even slow, miserable, mechanical nanocomputers-will give us practical machines with a trillion times the power of today's computers (essentially by letting us package a trillion computers in a small space, without gobbling too much money or energy.) The consequences will again be revolutionary.

Physics: The known principles of physics are adequate for understanding molecules, materials, and cells, but not for understanding phenomena on a scale that would still be submicroscopic if atoms were the size of marbles. Nanotechnology can't help here directly, but it can provide manufacturing facilities that will make huge particle accelerators economical, where today they strain national budgets.

More generally, nanotechnology will help science wherever precision and fine details are important. Science frequently proceeds by trying small variations in almost identical experiments, comparing the results. This will be easier when molecular manufacturing can make two objects that are identical, molecule by molecule. In some areas, today's techniques are not only crude, but destructive. Archaeological sites are unique records of the human past, but today's techniques throw away most information during the dig, by accident. Future archaeologists, able to sift soil not speck by speck but molecule by molecule, will be grateful indeed to those archaeologists who today leave some ground undisturbed.

Of all the areas where the ability to manufacture new tools is important, medicine is perhaps the greatest. The human body is intricate, and that intricacy extends beyond the range of human vision, beyond microscopic imaging, down to the molecular scale. "Molecular medicine" is an increasingly popular term today, but medicine today has only the simplest of molecular tools. As biology uses nanoinstruments to learn about disease and health, we will learn the physical requirements for restoring and maintaining health. And with this knowledge will come the tools needed to satisfy those requirements-tools ranging from improved pharmaceuticals to devices able to repair cells and tissues through molecular surgery.

Advanced medicine will be among the most complex and difficult applications of nanotechnology. It will require great knowledge, but nanoinstruments will help gather this knowledge. It will pose great engineering challenges, but computers of trillionfold greater power will help meet those challenges. It will solve medical problems on which we spend billions of dollars today, in hopes of modest improvements.

Today, modern medicine often means an expensive way to prolong misery. Will nanomedicine be more of the same? Any reader over the age of, say, thirty knows how things start to go wrong: an ache here, a wrinkle there, the loss of an ability. Over the decades, the physical quality of life declines faster and faster-the limits of what the body can do become stricter-until the limits are those of a hospital bed. The healing abilities we have when young seem to fade away. Modern medical practice expends the bulk of its effort on such things as intensive care units, dragging out the last few years of life without restoring health.

Truly advanced medicine will be able to restore and supplement the youthful ability to heal. Its cost will depend on the cost of producing things more intricate than any we have seen before, the cost of producing computers, sensors, and the like by the trillions. To understand the prospects for medicine, like those for science and industry, we need to take a closer look at the cost of molecular manufacturing.

In earlier chapters, we have stepped forward and backward through time. The last step was a big one, leaping from small laboratory devices to the high-capacity industrial facility of the Desert Rose scenario. Our narrative crossed this gap in a single leap, but the world won't. To understand how nanotechnology might unfold, it makes sense to look at some of its easier and more difficult applications. The result won't be a timetable, or even a series of milestones, but it should give a better picture of what we can expect as nanotechnology develops from simple, crude, costly beginnings to a state of greater sophistication and lower cost.

Molecular manufacturing will make better products possible. We're likely to see some early applications in at least two areas: stronger materials and faster computers. Strong materials are simple, and will be hard to pass up. Computers are more complex, but the payoff will be enormous.

REAL ACCOUNT

Some costs apply to a kind of product, regardless of how many copies of it are made: these include design costs, technology licensing costs, regulatory approval costs, and the like. Other costs apply to each unit of a product: these include the costs of labor, energy, raw materials, production equipment, production sites, insurance, and waste disposal. The per-kind costs can become very low if production runs are large. If these costs stay high, it will be because people prefer new products for their new benefits, despite the cost-hardly cause for complaint. The more basic and easier to analyze costs are per-unit costs. A picture to keep in mind here is of Desert Rose Industries, where molecular machinery does most of the work, and where products are made from parts that are ultimately made from simple chemical substances. Let's consider some cost components.

***Energy*:** Manufacturing at the molecular scale need not use a lot of energy. Plants build billions of tons of highly patterned material every year using available solar energy. Molecular manufacturing can be efficient, in the sense that the energy needed to build a block of product should be comparable to the energy released in burning an equivalent mass of wood or coal. If this energy were supplied as electricity at today's costs, the energy cost of manufacturing would be something like a dollar per kilogram. We'll return to the cost of energy later.

***Raw Materials*:** Molecular manufacturing won't need exotic materials as inputs. Plain bulk chemicals will suffice, and this means materials no more exotic than the fuels and feedstocks that are, for now, derived from petroleum and biomass-gasoline, methanol, ammonia, and hydrogen. These typically cost tens of cents per kilogram. If bizarre compounds are used, they can be made internally. Rare elements could be avoided, but might be useful in trace amounts. The total quantity of raw materials consumed will be smaller than in conventional manufacturing processes because less will be wasted.

Capital Equipment and Maintenance: As we saw in the Desert Rose scenario, molecular manufacturing can be used to build all of the equipment needed for molecular manufacturing. It seems that this equipment-everything from large vats to submicroscopic special-purpose assemblers-can be reasonably durable, lasting for months or years before being recycled and replaced. If the equipment were to cost dollars per kilogram, and produce many thousands of kilograms of product in its life, the cost of the equipment would add little to the cost of the product.

Waste Disposal: Today's manufacturing waste is dumped into the air, water, and landfills. There need be no such waste with molecular manufacturing. Excess materials of the kind now spewed into the environment could instead be completely recycled internally, or could emerge from the manufacturing process in pure form, ready for use in some other process. In an advanced process, the only wastes would be leftover atoms resulting from a bad mix of raw materials. Most of these leftover atoms would be ordinary minerals and simple gases like oxygen, the main "waste" from the molecular machinery of plants. Molecular manufacturing produces no new elements-if arsenic comes out, arsenic must have gone in, and the process isn't to blame for its existence. Any intrinsically toxic materials of this sort can at least be put in the safest form we can devise for disposal. One option would be to chemically bond it into a stable mineral and put it back where it came from.

Labor: Once a plant is operating, it should require little human labor (what people do with their time will change, unless factories are kept running as bizarre hobbies). Desert Rose Industries was run by two people, yet was described as producing large quantities of varied goods. The basic molecular-scale operations of manufacturing have to be automated, since they are too small for people to work on. The other operations are fairly simple and can be aided by equipment for handling materials and information.

Space: Even a manufacturing plant based on nanotechnology takes up room. It would, however, be more compact than familiar manufacturing plants, and could be built in some out-of-the-way place with inexpensive land. These costs should be small by today's standards.

Insurance: This cost will depend on the state of the law, but some comparisons can be made. Improved sensors and alarms could be made integral parts of products; these should lower fire and theft premiums. Product liability costs should be reduced by safer, more reliable products. Employee injury rates will be reduced by having less labor input. Still, the legal system in the United States has shown a disturbing tendency to block every new risk, however small, even when this forces people to keep suffering old risks, which are sometimes huge. (The supply of lifesaving vaccines has been threatened in just this way.) When this happens, we kill anonymous people in the name of safety. If this behavior raises insurance premiums in a perverse way, it could discourage a shift to safer manufacturing technologies. Since such costs can grow or shrink independent of the real world of engineering and human welfare, they are beyond our ability to estimate.

Sales, Distribution, Training: These costs will depend on the product: Is it as common as potatoes, and as simple to use? Or is it rare and complex, so that determining what you need, where to get it, and how to use it are the main problems? These service costs are real but can be distinguished from costs of the thing itself. To summarize, molecular manufacturing should eventually lead to lower costs. The initial expense of developing the technology and specific products will be substantial, but the cost of production can be low. Energy costs (at present prices) and materials costs (ditto) would be significant, but not enormous. They were quoted on a per-kilogram basis, but nanotechnological products, being made of superior materials, will often weigh only a fraction of what familiar products do. (Ballast, were it needed, will be dirt-cheap.) Equipment costs, land costs, waste-disposal costs, and labor costs can be low by the very nature of the technology.

Costs of design, regulation, and insurance will depend strongly on human tastes and are beyond predicting. Basic products, like clothing and housing, can become inexpensive unless we do something to keep them costly. As the cost of improved safety falls, there will be less reason to accept unsafe products. Molecular manufacturing uses processes as controlled and efficient as the molecular processes in plants. Its products could be as inexpensive as potatoes. This may sound to good to be true (and there are downsides, as we'll discuss), but why shouldn't it be true? Shouldn't we expect large changes to come with the replacement of modern technology?

PROGRESS OF NANOTECHNOLOGY

A basic question about nanotechnology is, "When will it be achieved?" The answer is simple: No one knows. How molecular machines will behave is a matter for calculation, but how long it will take us to develop them is a separate issue. Technology timetables can't be calculated from the laws of nature, they can only be guessed at. In this chapter, we examine different paths to nanotechnology, hear what some of the pioneers have to say, and describe the progress already made. This will not answer our basic question, but it will educate our guesses.

Molecular nanotechnology could be developed in any of several basically different ways. Each of these basic alternatives itself includes further alternatives. Researchers will be asking, "How can we make the fastest progress?" To understand the answers they may come to, we need to ask the same question here, adopting (for the moment) a gung-ho, let's-go, how-do-we-get-the-job-done? attitude. We give some of the researchers' answers in their own words.

Like "When will it be achieved?", this is a basic question with an answer beyond calculation. Here, though, the answer seems fairly clear. Throughout history, people have worked to achieve better control of matter, to convince atoms to do what we want them to do. This has gone on since before people learned that atoms exist, and has accelerated ever since. Although different industries use different materials and different tools and methods, the basic aim is always the same. They seek to make better things, and make them more consistently, and that means better control of the structure of matter. From this perspective, nanotechnology is just the next, natural step in a progression that has been under way for millennia.

Consider the compact discs now replacing older stereo records: both the old and the new technologies stamp patterns into plastic, but for CDs, the bumps on the stamping surface are only about 130 by 600 nanometers in size, versus 100,000 nanometers or so for the width of the groove on an old-style record. Or look at a personal computer.

John Foster, a physicist at IBM's Almaden Research Center, points to a hard disk and says that "inside that box are a bunch of whirring disks, and every one of those disks has got a metal layer where the information is stored. The last thing on top of the metal layer is a monolayer that's the lubricant between the disk and the head that flies over it. The monolayer is not fifteen angstroms [15 angstroms = 1.5 nanometers] and it's not three, because fifteen won't work and neither will three. So it has to be ten plus or minus a few angstroms. This is definitely working in the nanometer regime. We're at that level: We ship it every day and make money on it every day."

The transistors on computer chips are heading down in size on an exponential curve. Foster's colleague at IBM, Patrick Arnett, expects the trend to continue: "If you stay on that curve, then you end up at the atomic scale at 2020 or so. That's the nature of technology now. You expect to follow that curve as far as you can go."

The trend is clear, and at least some of the results can be foreseen, but the precise path and timetable for the development of nanotechnology is unpredictable. This unpredictability goes to the heart of important questions:

"How will this technology be developed? Who will do it? Where? When? In ten years? Fifty? A hundred? Will this happen in my lifetime?" The answers will depend on what people do with their time and resources, which in turn will depend on what goals they think are most promising. Human attitudes, understanding, and goals will make all the difference.

MORE FAMILIAR KINDS OF ENGINEERING

Manufacturers and architects know that designs for new products and buildings are best done on a computer, by computer-aided design (CAD). The new molecular design software can be called molecular CAD, and in its forefront are researchers such as Jay Ponder of the Yale University Department of Molecular Biophysics and Biochemistry. Ponder explains that "There's a strong link between what molecular designers are doing and what architects do. Michael Ward of Du Pont is designing a set of building blocks for a Tinker Toy set so that you can build larger structures. That's exactly what we're doing with molecular modeling techniques.

"All the design and mechanical engineering principles that apply to building a skyscraper or a bridge apply to molecular architecture as well. If you're building a bridge, you're going to model it and see how many trucks can be on the bridge at the same time without it collapsing, what kind of forces you're going to apply to it, whether it can stand up to an earthquake.

"And the same process goes on in molecular design: You're designing pieces and then analyzing the stresses and forces and how they will change and perturb the structure. It's exactly the same as designing and building a building, or analyzing the stresses on any macroscale structure. I think it's important to get people to think in those terms.

"The molecular designer has to be creative in the same way that an architect has to be creative in designing a building. When people are looking at the interior of a protein structure and trying to redesign it to create a space that will have a particular function, such as binding to particular molecules, that's like designing a room to use as a dining room-one that will fit certain sizes of tables and certain numbers of guests. It's the same thing in both cases: You have to design a space for a function."

Ponder combines chemistry and computer science with an overall engineering approach: "I'm kind of a hybrid. I spend about half my time doing experiments and about half my time writing computer programs and doing computational work. In the laboratory, I create or design molecules to test some of the computational ideas. So I'm at the interface." The engineering perspective helps in thinking about where molecular research can lead: "Even though with nanotechnology we're at the nanometer scale, the structures are still big enough that an awful lot of things are classical. Again, it's really like building bridges-very small bridges. And so there are many almost standard mechanical-engineering techniques for architecture and building structures, such as stress analysis, that apply."

Getting to nanotechnology will require the work of experts in differing fields: chemists, who are learning how to make molecular machines; computer scientists, who are building the needed design tools; and perhaps STM and AFM experts, who can provide tools for molecular positioning. To make progress, however, these experts must do more than just work, they must work together. Because nanotechnology is inherently interdisciplinary, countries that draw hard lines between their academic disciplines, as the United States does, will find that their researchers have difficulty communicating and cooperating.

In chemistry today, a half-dozen researchers aided by a few tens of students and technicians is considered a large team. In aerospace engineering, enormous tasks like reaching the Moon or building a new airliner are broken down into tasks that are within the reach of small teams. All these small teams work together, forming a large team that may consist of thousands of engineers aided by many thousands of technicians. If chemistry is to move in the direction of molecular systems engineering, chemists will need to take at least a few steps in this direction.

In engineering, everyone knows that designing a rocket will require skills from many disciplines. Some engineers know structures, others know pumps, combustion, electronics, software, aerodynamics, control theory, and so on and so forth down a long list of disciplines. Engineering managers know how to bring different disciplines together to build systems.

In academic science, interdisciplinary work is productive and praised, but is relatively rare. Scientists don't need to cooperate to have their results fit together: they are all describing different parts of the same thing-nature-so in the long run, their results tend to come together into a single picture. Engineering, however, is different. Because it is more creative (it actually creates complex things), it demands more attention to teamwork. If the finished parts are going to work together, they must be developed by groups that share a common picture of what each part must accomplish. Engineers in different disciplines are forced to communicate; the challenge of management and team-building is to make that communication happen. This will apply to engineering molecular systems as much as it does to engineering computers, cars, aircraft, or factories.

Jay Ponder suggests that it's a question of perspective. "It's all a matter of what's perceived to be important by the different groups that have to come together to make this work: the chemists doing their bit and the computational people doing their bit. People have to come together and see the big picture. There are people who try to bridge the gaps, but they are rare compared to the people who just work in their own specialty." Progress toward nanotechnology will continue, and as it does, researchers trained as chemists, physicists, and the like will learn to talk to one another to solve new problems. They will either learn to think like engineers and work in teams, or they will be eclipsed by colleagues who do.

With all these problems, the advance toward nanotechnology steadily continues. Industry must gain ever-better control of matter to stay competitive in the world marketplace. The STM, protein engineering, and much of chemistry are driven by commercial imperatives. Focused efforts would yield faster advances, yet even without a clear focus, advances in this direction have an air of inevitability. As Bill DeGrado observes, "We really do have the tools. Experience has shown that when you have the analytic and synthetic tools to do things, in the end science goes ahead and does them-because they are doable."

Jay Ponder agrees: "Over the next few years, you're going to see slow evolutionary advances coming from people tinkering with molecular structures and figuring out their principles. People are going to work on a particular problem because they see some application for it or because they got grant funding for it. And in the process of doing something like improving a laundry detergent's ability to clean protein stains, Proctor and Gamble is going to help work out the principles for how to increase molecular stability, and to design spaces inside the molecules." For a variety of reasons, Japan's contribution to nanotechnology research promises to be excellent. While the United States has generally pursued researching this area with little sense of long-term direction, it appears that Japan has begun to take a more focused approach. Researchers there already have clear ideas about molecular machines-about what might work and what probably won't. Japanese researchers are accustomed to a higher level of interdisciplinary contact and engineering emphasis than are Americans. In the United States, we prize "basic science," often calling it "pure science," as if to imply that practical applications are a form of impurity. Japan instead emphasizes "basic technology."

Nanotechnology is a basic technology, and the Japanese recognize it as such. Recent changes at the Tokyo Institute of Technology-Japan's equivalent of MIT-reflect their views of promising directions for future research. For many decades, Tokyo Tech has had two major divisions: a Faculty of Science and a Faculty of Engineering. To these is now being added a Faculty of Bioscience and Biotechnology, to consist of four departments: a Department of Bioscience, a Department of Bioengineering, a Department of Biomolecular Engineering, and what is termed a "Department of Biostructure." The creation of a new faculty in a major Japanese university is a rare event. What U.S. university has a department explicitly devoted to molecular engineering? Japan has both the departments at Tokyo Tech and Kyoto University's recently established Department of Molecular Engineering.

Japan's Institute for Physical and Chemical Research (RIKEN) has broad-based interdisciplinary strength. Hiroyuki Sasabe, head of the Frontier Materials Research Program at RIKEN, notes that the institute has expertise in organic synthesis, protein engineering, and STM technology. Sasabe says that his laboratory may need a molecular manipulator of the sort described in the next chapter to accomplish its goals in molecular engineering. Research consortia in Japan are also moving toward nanotechnology. The Exploratory Research for Advanced Technology Organization (ERATO) sponsors many three-to-five year projects in parallel, each with a specific goal. Consider the work in progress:

- o Yoshida Nanomechanism Project
- o Hotani Molecular Dynamic Assembly Project
- o Kunitake Molecular Architecture Project
- o Nagayama Protein Array Project
- o Aono Atomcraft Project

These focus on different aspects of gaining control over matter at the atomic level. The Nagayama Protein Array Project aims to use proteins as engineering materials to move toward making new molecular devices. The Aono Atomcraft Project does not involve nuclear power-as its translation might imply-but is instead an interdisciplinary effort to use an STM to arrange matter on the atomic scale.

At some point, work on nanotechnology must move beyond spin-offs from other fields and undertake the design and construction of molecular machinery. This shift from opportunistic science to organized engineering requires a change in attitude. In this, Japan leads the United States.

Molecular nanotechnology will emerge step by step. Major milestones, such as the engineering of proteins and the positioning of individual atoms, have already been passed. To get a sense of the likely pace of developments, we need to look at how various trends fit together.

Computer-based molecular-modeling tools are spawning computer-aided design tools. These will grow more capable. The underlying technology base-computer hardware-has for decades been improving in price and performance on a steeply rising curve, which is generally expected to continue for many years. These advances are quite independent of progress in molecular engineering, but they make molecular engineering easier, speeding advances. Computer models of molecular machines are beginning to appear, and these will whet the appetites of researchers.

Progress in engineering molecular machines, whether using proximal probes or self-assembly, will eventually achieve striking successes; the objectives of research in Japan will begin to draw serious attention; understanding of the long-term promise of molecular engineering will become more widespread.

Some combination of these developments will eventually lead to a serious, public appraisal of what these technologies can achieve-and then the world of opinion, funding, and research fashion will change. Before, advances will be steady but haphazard; afterward, advances will be driven with the energy that flows into major commercial, military, and medical research programs, because nanotechnology will be recognized as furthering major commercial, military, and medical goals. The timing of subsequent events depends largely on when this threshold of serious attention is reached.

In making time estimates, people are prone to assume that a large change must take a long time. Most do, but not all. Pocket calculators had a dramatic effect on the slide-rule industry: they replaced it.

The speed of this change caught the slide rule moguls by surprise, but the pace of progress in electronics didn't slow down merely to suit their expectations. One can argue that nanotechnology will be developed fast: many countries and companies will be competing to get there first. They will be driven onward both by the immense expected benefits-in many areas, including medicine and the environment-as well as by potential military applications. That is a powerful combination of motives, and competition is a powerful accelerator.

A counterargument, though, suggests that development will be slow: anyone who has done anything of significance in the real world of technology-doing a scientific experiment, writing a computer program, bringing a new product to market-knows that these goals take longer than expected. Indeed, Hofstadter's Law states that projects take longer than expected, even when Hofstadter's Law is taken into account. This principle is a good guide for the short term, and for a single project.

The situation differs, though, when many different approaches are being explored by many different groups over a period of years. Most projects may take longer than expected, but with many teams trying many approaches, one approach may prove faster than expected. The winner of a race is always faster than the average runner. John Walker notes, "The remarkable thing about molecular engineering is that it looks like there are many different ways to get there and, at the moment, rapid progress is being made along every path-all at the same time."

Also, technology development is like a race run over an unmapped course. When the first runners reach the top of a hill, they may see a shortcut. A trailing runner may decide to crash off into the bushes, and stumble across a bicycle and a paved road. The progress of technology is seldom predictable because progress often reveals new directions. How close we are to goal depends on whether technological advances are a constant pace of accelerating. In this diagram, the dashed line represents the current level of technology, and the large dot in the upper right represents a goal such as nanotechnology. With a straight-line advance, it's easier to estimate how far away a goal is. With an accelerating advance, a goal can be reached with little warning.

So how can we estimate a date for the arrival of nanotechnology? It's safest to take a cautious approach: When anticipating benefits, assume it's far off; when preparing for potential problems, assume it's right around the corner. The old folk saying applies: Hope for the best, prepare for the worst. Any dates assigned to "far off" and "right around the corner" can be no better than educated guesses-molecular behavior can be calculated, but not technology timetables of this sort. With those caveats, we would estimate that general-purpose molecular assemblers will likely be developed in the early decades of the twenty-first century, perhaps in the first.

John Walker, whose technological foresight has led Autodesk from start-up to a dominant role in its industry, points out that not long ago, "Many visionaries intimately familiar with the developments of silicon technology still forecast it would take between twenty and fifty years before molecular engineering became a reality. This is well beyond the planning horizon of most companies. But recently, everything has begun to change." Based on the new developments, Walker places his bet: "Current progress suggests the revolution may happen within this decade, perhaps starting within five years.

4

Molecular Manufacturing and the Developing World

Molecular manufacturing (MM) can solve many of the world's current problems. For example, water shortage is a serious and growing problem. Most water is used for industry and agriculture; both of these requirements would be greatly reduced by products made by molecular manufacturing. Infectious disease is a continuing scourge in many parts of the world.

Simple products like pipes, filters, and mosquito nets can greatly reduce this problem. Information and communication are valuable, but lacking in many places. Computers and display devices would become stunningly cheap. Electrical power is still not available in many areas.

The efficient, cheap building of light, strong structures, electrical equipment, and power storage devices would allow the use of solar thermal power as a primary and abundant energy source. Environmental degradation is a serious problem worldwide.

High-tech products can allow people to live with much less environmental impact. Many areas of the world cannot rapidly bootstrap a 20th century manufacturing infrastructure. Molecular manufacturing technology can be self-contained and clean; a single packing crate or suitcase could contain all equipment required for a village-scale industrial revolution. Finally, MM will provide cheap and advanced equipment for medical research and health care, making improved medicine widely available. Much social unrest can be traced directly to material poverty, ill health, and ignorance. MM can contribute to great reductions in all of these problems, and in the associated human suffering.

Despite limited literature discussing the global implications of molecular manufacturing (MM), the seeds for certain key debates are starting to be sown. They essentially mirror those presented for current and near-term nanotechnology: for what purposes will the technology be developed and used, by whom will it be created and owned, what is the nature of the risks it will bring, and what kind of impact will it have upon the global economy and developing world?

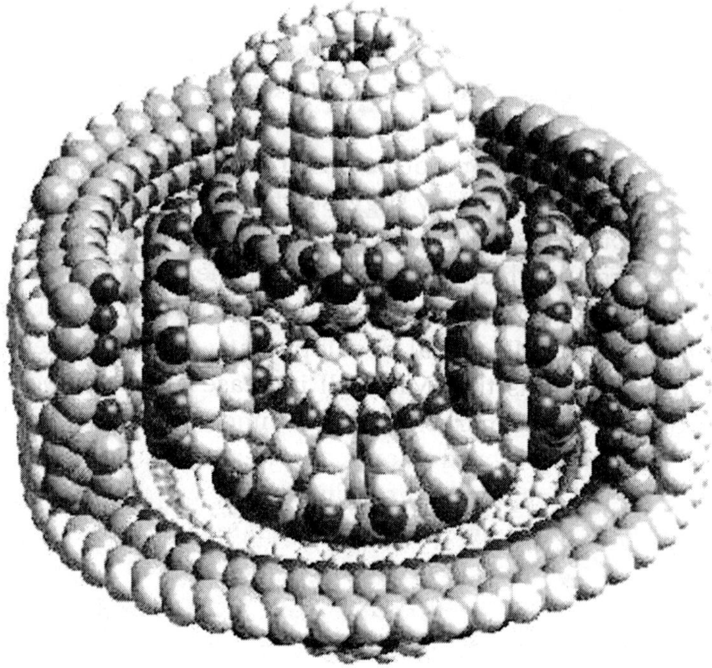

Figure: ***Molecular Manufacturing***

The realisation of MM's central goals would undoubtedly lead to the most concentrated technological 'tsunami' ever witnessed. The unpredictable nature of such a revolution makes answering the previous questions all the more difficult. However, just as biotechnology is commonly used as a yardstick for nanotechnology evaluations (given the continuity of many social concerns and relationship between existing biotechnology capabilities and potential to develop nanotechnology), we argue here that the best way to set up the MM debate, in terms of its impact on developing countries, may be to look at developments and trends in nanotechnology. Central to such analysis is addressing the ways in which nanotechnology creates new possibilities for developing countries in terms of access to technology, potential benefits, risks, and shifting views of science and technology, as well as the imposing of new demands in terms of infrastructure and approaches to science.

While many of the issues MM faces may be similar to those presently developing with nanotechnology, MM offers a revolution of a starkly different magnitude. However, MM still faces an 'identity crisis' in the developed world, and an 'identity absence' in the developing world. This has been further hampered by authors and academics who, in writing articles and papers concerning nanotechnology's impact upon the developing world, mix the two terms and confuse references to the relevant potential impacts. Coupled with hype surrounding 'grey goo', a poor foundation for international discussions in which to include MM has evolved. There need to be consistent efforts by those writing in the area to distinguish the kind of nanotechnology to which they are referring.

However, while the scope of potential impact differs greatly between nanotechnology and MM, history suggests there will be universal patterns in terms of its distribution. Whether MM is developed to either a limited or a full capacity, it is more than likely the majority of the world will never reap its benefits and perhaps even fewer will know or understand its potential.

The location of MM's initial development will play a key role in the resolution of suggestions that MM will increase global corporate control, eradicate natural resource markets to the detriment of developing countries or reduce global inequalities and allow countries to 'leap-frog' the industrial revolution. Some suggest that industrialised countries are more fertile ground for its development, while others believe China and India are in a more likely position than the United States of America and European Union to be the initial producers. However, as McCarthy notes, "just possessing the technology does not mean that a given state will be able to use it to full advantage". Countries will require an ability to integrate and respond to MM in order to exploit its potential.

In more recent times, debates have extended to consider MM's initial development by both small and big research groups, within private and public spheres, by structured and unstructured research groups, as well as by 'rogue' and 'peaceful' nations.

Assessing nanotechnology, we notice an increasing 'nano-divide' in terms of national research levels and funding, as much amongst the developing countries themselves as between the developed and the developing world. This highlights the limitations of discussing the developing world as if it was an homogeneous grouping. Emerging technology affects regions, countries, populations and communities in vastly disparate ways. Already, 62 countries are engaging with nanotechnology on a national level. Nineteen of them are classified as 'developing' but none of these are from the group of countries classified as 'least developed'. However, via nanotechnology we may be witness to a new form of inexpensive access to niche R&D markets, as demonstrated by development of national nanotechnology programs in countries such as Costa Rica.

Furthermore, when spending is adjusted for purchasing-power parity, Chinese government nanotechnology funding ranks second, internationally, behind the United States of America. Nonetheless, funding (on an absolute basis) remains heavily focussed towards programs in industrialised countries of the European Union, the United States of America and Japan.

It appears that smaller national programs will require clear, specific strategic road maps that target niche research and consider significant international collaboration if they wish to be part of the 'nanotechnology revolution'.

Initial applications in nanotechnology, such as in cosmetics, sporting apparatus and clothing, have exposed early indicators as to the nature of research orientation. Increasing levels of private sector patent concentration within the developed world, promoting 'broad brush' patents, have added to developing country concerns. By the time MM arrives, corporate control in areas such as the life sciences will be so strong and patents and other forms of technological control so all-encompassing, that the ability to replicate a multi-functional theranostic kit cheaply and rapidly may have little or no bearing on its ability to reach those in greatest need.

Some have suggested that 'economic abundance' resulting from the realisation of MM will negate such a trend. MacGillivray disagrees, believing that with MM "the economics of production will change… human nature won't." History also disagrees. While economic theory is based on scarcity, the current world is already one of abundance. We 'solved' the 'production issue' long ago. Given the world's current population, we do not need greater production to feed such a number.

Rather, we require more equitable distribution of resources, reductions in consumption and the recognition that true 'equality' (at Western standards of living) is not sustainable. As the 2002 United Nations' Human Development report notes, if the per capita energy consumption of developing countries were to rise to even half of that of the advanced industrial economies, the energy reserves of this finite planet soon would be exhausted.

Despite the orientation of early applications, a number of groups remain hopeful that nanotechnology can be 'appropriate' and used to fulfil the Millennium Development Goals. Similarly, 'nanofactories' have been touted as 'appropriate technology' on the basis of their potential to reduce skilled labour and supporting infrastructure requirements. However, it depends whether a nanofactory would reduce skilled or basic labour. If the latter, then it fails Schumacher's criteria for an appropriate technology.

Furthermore, assessments of appropriateness must include assessments of risk. As with nanotechnology, suggestions have been made that MM could promote world peace while others have cautioned that it may, initially, prove quite dangerous. Consistently, reports are finding that research into nanotechnology risk is inadequate and under-funded. With MM offering less chance to both predict and react to issues of risk, the key must be to focus on global capacity building in areas of risk assessment and risk management.

The problem with developed country discussions of nanotechnology for the developing world is that they often don't progress beyond the identification of so-called 'appropriate' applications. Very few people have even arrived at an acceptance that there are 'appropriate applications', full stop. MM continues to struggle for mainstream acceptance and it is, perhaps, more likely that its fruition will come via a gradual extension of nanoscale capabilities, rather than an immediate jump to MM.

In terms of capacity requirements, nanotechnology is changing both the way science is conducted within research teams and promoted to the public. As we move to a scale unable to be viewed by a standard microscope, and less likely to be comprehended, environmental and human health risk become increasingly important issues with which to engage the public. Early commentators have suggested that nanotechnology will reinforce the lessons from the GM-debate: that public participation is crucial to acceptance but 'engineering consent' is not a sustainable means for widespread acceptance and development. In my experience with researchers in Thailand, India and Australia, there is a strong recognition that capacity development in nanotechnology must incorporate ethicists and legal experts in terms of analysing national regulatory frameworks, maximising competitive advantage and contributing to public dialogue. Finally, nanotechnology is demanding and creating a more interdisciplinary view of science at the research level. Since MM will most probably enter in a climate shaped by nanotechnology's legacy in these areas, it is important that such considerations extend to MM debates.

All these considerations lead me to make two suggestions.

The first suggestion is that we work towards a truly international nanotechnology conference some time around 2010. Such a conference would ideally occur within the United Nations framework and involve official country representatives. It would be of great benefit to the conference, in terms of productivity, if countries submitted 'white papers' prior to the commencement of the proceedings, outlining national capacities, strategies and proposed niche markets in nanotechnology. Such a meeting could also incorporate discussion on the proposed International Convention for the Evaluation of New Technologies and gauge reception of and progress towards MM. The Global Nanotechnology Network as well as the Asia Nano Forum provide good examples and existing infrastructure for any efforts to develop international cooperation and consensus.

The second suggestion is that, replicating a method such as the United Kingdom's Royal Academy of Engineering 'upstreaming' approach, civil society groups should seek to increase nanotechnology and MM exposure at international 'grassroots' events such as the World Social Forum. Such efforts hopefully would pave the way for greater developing country engagement in academic writing and scholarship pertaining to these areas.

Despite significant differences in the potential capabilities of nanotechnology and MM, nanotechnology may offer the greatest insights and means by which to influence the future impact of MM with respect to the developing world. Initially, a clear distinction must be made between nanotechnology and MM. Current nanotechnology indicators point towards increasing concern for the developing world in terms of barriers to technology access, inadequate research into environmental and human health risks, and significant demands in terms of the capacity to respond to and develop nanotechnology. Any study of MM and its potential impacts upon the developing world will gain from an appreciation of the relevant context and developments surrounding nanotechnology.

TECHNICAL BASIS OF MOLECULAR MANUFACTURING

At its foundation, molecular manufacturing works by doing a few precise fabrication operations, very rapidly, at the molecular level, under computer control. It can thus be viewed as a combination of mechanical engineering and chemistry, with some additional help from rapid prototyping, automated assembly, and related fields of research. Atoms and inter-atomic bonds are completely precise: every atom of a type is identical to every other, and there are only a few types. Barring an identifiable error in fabrication, two molecules manufactured according to the same blueprint will be identical in structure and shape (with transient variations of predictable scale due to thermal noise and other known physical effects). This consistency will allow fully automated fabrication. Computer controlled addition of molecular fragments, creating a few well-characterized bond types in a multitude of selected locations, will enable a vast range of components to be built with extremely high reliability. Building with reliable components, higher levels of structure can retain the same predictability and engineerability.

A fundamental "scaling law" of physics is that small systems operate faster than large systems. Moving at moderate speed over tiny distances, a nanoscale fabrication system could perform many millions of operations per second, creating products of its own mass and complexity in hours or even minutes. Along with faster operation comes higher power density, again proportional to the shrinkage: nanoscale machines might be a million times more compact than today's technology.

Computers would shrink even more profoundly, and non-electronic technologies already analyzed could dissipate enough less power to make the shrinkage feasible. Although individual nanoscale machines would have small capacity, massive arrays could work together; it appears that gram-scale computer and motor systems, and ton-scale manufacturing systems, preserving nanoscale performance levels, can be built without running afoul of scaling laws or other architectural constraints including cooling. Thus, products will be buildable in a wide range of sizes.

A complete list of advantages and capabilities of molecularly manufactured products, much less an analysis of the physical basis of the advantages, would be beyond the scope of this paper. But several additional advantages should be noted.

Precisely fabricated covalent materials will be much stronger than materials formed by today's imprecise manufacturing processes. Precise, well-designed, covalently structured bearings should suffer neither from wear nor from static friction (stiction). Carbon can be an excellent conductor, an excellent insulator, or a semiconductor, allowing a wide range of electrical and electronic devices to be built in-place by a molecular manufacturing system.

Development of Molecular Manufacturing

Although its capabilities will be far-reaching, the development of molecular manufacturing may require a surprisingly small effort. A finite, and possibly small, number of deposition reactions may suffice to build molecular structures with programmable shape—and therefore, diverse and engineerable function. High-level architectures for integrated kilogram-scale arrays of nanoscale manufacturing systems have already been worked out in some detail. Current-day tools are already able to remove and deposit atoms from selected locations in covalent solids. Engineering of protein and other biopolymers is another pathway to molecularly precise fabrication of engineered nanosystem components. Analysis tools, both physical and theoretical, are developing rapidly.

As a general rule, nanoscale research and development capabilities are advancing in proportion to Moore's Law—even faster in some cases. Conceptual barriers to developing molecular manufacturing systems are also falling rapidly. It seems likely that within a few years, a program to develop a nanofactory will be launched; some observers believe that one or more covert programs may already have been launched.

It also seems likely that, within a few years of the first success, the cost of developing an independent capability will have dropped to the point where relatively small groups can tackle the project. Without stringent and widespread restrictions on technology, it most likely will not be possible to prevent the development of multiple molecular manufacturing systems with diverse owners.

Products of Molecular Manufacturing

All exploratory engineering in the field to date has pointed to the same set of conclusions about molecular manufacturing-built products:

1. Manufacturing systems can build more manufacturing systems.
2. Small products can be extremely compact.
3. Human-scale products can be extremely inexpensive and lightweight.
4. Large products can be astonishingly powerful.

If a self-contained manufacturing system can be its own product, then manufacturing systems can be inexpensive—even non-scarce. Product cost can approach the cost of the feedstock and energy required to make it (plus licensing and regulatory overhead). Although molecular manufacturing systems will be extremely portable, most products will not include one—it will be more efficient to manufacture at a dedicated facility with installed feedstock, energy, and cooling supplies. The feature size of nanosystems will probably be about 1 nanometer (nm), implying a million features in a bacteria-sized object, a billion features per cubic micron, or a trillion features in the volume of a ten-micron human cell. A million features is enough to implement a simple CPU, along with sensors, actuators, power supply, and supporting structure. Thus, the smallest robots may be bacteria-sized, with all the scaling law advantages that implies, and a medical system (or weapon system based thereon) could be able to interact with cells and even sub-cellular structures on an equal footing. As a general rule of thumb, human-scale products may be expected to be 100-1000 times lighter than today's versions. Covalent carbon-based materials such as buckytubes should be at least 100 times stronger than steel, and materials could be used more efficiently with more elegant construction techniques. Active components will shrink even more. (Of course, inconveniently light products could be ballasted with water.)

Large nanofactories could build very large products, from spacecraft to particle accelerators. Large products, like smaller ones, could benefit from stronger materials and from active systems that are quite compact. Nanofactories should scale to at least ton-per-hour production rates for integrated products, though this might require massive cooling capacity depending on the sophistication of the nanofactory design.

Possible Weapons Systems

The smallest systems may not be actual weapons, but computer platforms for sensing and surveillance. Such platforms could be micron-scale. The power requirement of a 1-MIPS computer might be on the order of 10-100 pW; at that rate, a cubic micron of fuel might last for 100-1000 seconds. The computer itself would occupy approximately one cubic micron.

Very small devices could deliver fatal quantities of toxins to unprotected humans. Even the smallest ballistic projectiles (bullets) could contain supercomputers, sensors, and avionics sufficient to guide them to targets with great accuracy. Flying devices could be quite small. It should be noted that small devices could benefit from a process of automated design tuning; milligram-scale devices could be built by the millions, with slight variations in each design, and the best designs used as the basis for the next "generation" of improvements; this could enable, for example, UAV's in the laminar regime to be developed without a full understanding of the relevant physics. The possibility of rapid design is far more general than this, and will be discussed below.

The line between bullets, missiles, aircraft, and spacecraft would blur. With lightweight motors and inexpensive manufacturing, a vehicle could contain a number of different disposable propulsion systems for different flight regimes. A "briefcase to orbit" system appears feasible, though such a small device might have to fly slowly to conserve fuel until it reached the upper atmosphere. It might be feasible to use 1 kg of airframe (largely discarded) and 20 kg of fuel (not counting oxidizer) to place 1 kg into orbit; some of the fuel would be used to gather and liquify oxygen in the upper atmosphere for the rocket portion of its flight. (Engineering studies have not yet been done for such a device, and it might require somewhat more fuel than stated here.)

Advanced construction could produce novel energy-absorbing materials involving high-friction mechanical slippage under high stress via micro-or nano-scale mechanical components. In effect, every molecule would be a shock absorber, and the material could probably absorb mechanical energy until it was destroyed by heat.

New kinds of weapons might be developed more quickly with rapid inexpensive fabrication. Many classes of device will be buildable monolithically. For example, a new type of aircraft or even spacecraft might be tested an order of magnitude more rapidly and inexpensively, reducing the cost of failure and allowing further acceleration in schedule and more aggressive experimentation. Although materials and molecular structures would not encompass today's full range of manufactured substances, they could encompass many of the properties of those substances, especially mechanical and electrical properties. This may enable construction of weapons such as battlefield lasers, rail guns, and even more exotic technologies.

Passive armor certainly could not stop attacks from a rapid series of impacts by precisely targeted projectiles. However, armor could get a lot smarter, detecting incoming attacks and rapidly shifting to interpose material at the right point. There may be a continuum from self-reconfiguring armor, to armor that detaches parts of itself to hurl in the path of incoming attacks, to armor that consists of a detached cloud of semi-independent counterweapons.

A new class of weapon for wide-area destruction is kinetic impact from space. Small impactors would be slowed by the atmosphere, but medium-to-large asteroids, redirected onto a collision course, could destroy many square miles. The attack would be detectable far in advance, but asteroid deflection and destruction technology is not sufficiently advanced at this time to say whether a defender with comparable space capacity could avoid being struck, especially if the asteroid was defended by the attacker. Another class of space impactor is lightweight solar sails accelerated to a respectable fraction of light speed by passage near the sun. These could require massive amounts of inert shielding to stop; it is not clear whether or not the atmosphere would perform this function adequately. A hypothetical device often associated with molecular manufacturing is a small, uncontrolled, exponentially self-replicating system. However, a self-replicating system would not make a very good weapon. In popular conception, such a system could be built to use a wide range of feedstocks, deriving energy from oxidizing part of the material (or from ambient light), and converting the rest into duplicate systems. In practice, such flexibility would be quite difficult to achieve; however, a system using a few readily available chemicals and bypassing the rest might be able to replicate itself—though even the simplest such system would be extremely difficult to design. Although unrestrained replication of inorganic systems poses a theoretical risk of widespread biosphere destruction through competition for resources—the so-called "grey goo" threat—it seems unlikely that anyone would bother to develop grey goo as a weapon, even a doomsday deterrent. It would be more difficult to guide than a biological weapon. It would be slower than a device designed simply to disrupt the physical structure of its target. And it would be susceptible to detection and cleanup by the defenders.

Tactics

A detailed analysis of attack and defense is impossible at this point. It is not known whether sensor systems will be able to effectively detect and repel an encroachment by small, stealthy robotic systems; it should be noted that the smallest such systems might be smaller than a wavelength of visible light, making detection at a distance problematic. It is unknown whether armor will be able to stop the variety of penetrating objects and forces that could be directed at it. Semi-automated R&D may or may not produce new designs so quickly that the side with the better software will have an overwhelming advantage. The energy cost of construction has only been roughly estimated, and is uncertain within at least two orders of magnitude; active systems, including airframes for nano-built weapons, will probably be cost-effective in any case, but passive or static systems including armor may or may not be worth deploying.

Some things appear relatively certain. Unprotected humans, whether civilian or soldier, will be utterly vulnerable to nano-built weapons. In a scenario of interpenetrating forces, where a widespread physical perimeter cannot be established, humans on both sides can be killed at will unless protected at great expense and inconvenience. Even relatively primitive weapons such as hummingbird-sized flying guns with human target recognition and poisoned bullets could make an area unsurvivable without countermeasures; the weight of each gun platform would be well under one gram. Given the potential for both remote and semi-autonomous operation of advanced robotics and weapons, a force with a developed molecular manufacturing capability should have no need to field soldiers; this implies that battlefield death rates will be low to zero for such forces.

A concern commonly raised in discussions of nanotech weapons is the creation of new diseases. Molecular manufacturing seems likely to reduce the danger of this. Diseases act slowly and spread slowly. A sufficiently capable bio-sensor and diagnostic infrastructure should allow a very effective and responsive quarantine to be implemented. Rapid testing of newly manufactured treatment methods, perhaps combined with metabolism-slowing techniques to allow additional R&D time, could minimize disease even in infected persons

Despite the amazing power and flexibility of molecular manufactured devices, a lesson from World War I should not be forgotten: Dirt makes a surprisingly effective shield. It is possible that a worthwhile defensive tactic would be to embed an item to be protected deeply in earth or water. Without active defenses, which would also be hampered by the embedding material, this would be at best a delaying tactic. Information is likely to be a key determiner of military success. If, as seems likely, unexpected offense with unexpected weapons can overwhelm defense, then rapid detection and analysis of an attacker's weapons will be very important. Information-gathering systems are likely to survive more by stealth than by force, leading to a "spy vs. spy" game. To the extent that this involves destruction of opposing spy-bots, it is similar to the problem of defending against small-scale weapons. Note that except for the very smallest systems, the high functional density of molecularly constructed devices will frequently allow both weapon and sensor technology to be piggybacked on platforms primarily intended for other purposes.

It seems likely that, with the possible exception of a few small, fiercely defended volumes, a robotic battleground would consist of interpenetrated forces rather than defensive lines (or defensive walls). This implies that any non-active matter could be destroyed with little difficulty unless shielded heavily enough to outlast the battle.

Strategy

As implied above, a major strategy is to avoid putting soldiers on the battlefield via the use of autonomous or remotely operated weapons. Unfortunately, this implies that an enemy wanting to damage human resources will have to attack either civilian populations or people in leadership positions. To further darken the picture, civilian populations will be almost impossible to protect from a determined attack without maintaining a near-hermetic seal around their physical location. Since the resource cost of such a shield increases as the shield grows (and the vulnerability and unreliability probably also increase), this implies that civilians under threat will face severe physical restrictions from their own defenders.

A substantial variety of attack mechanisms will be available, including kinetic impact, cutting, sonic shock and pressure, plasma, electromagnetic beam, electromagnetic jamming and EMP, heat, toxic or destructive chemicals, and perhaps more exotic technologies such as particle beam and relativistic projectile. A variety of defensive techniques will be available, including camouflage, small size, physical avoidance of attack, interposing of sacrificial mass, jamming or hacking of enemy sensors and computers, and preemptive strike. Many of these offensive and defensive techniques will be available to devices across a wide range of sizes. As explored above, development of new weapon systems may be quite rapid, especially if automated or semi-automated design is employed.

In addition to the variety of physical modes of attack and defense, the cyber sphere will become an increasingly important and complex battleground, as weapon systems increasingly depend on networking and computer control. It remains to be seen whether a major electronic security breach might destroy one side's military capacity, but with increasing system complexity, such an occurrence cannot be ruled out.

Depending on what is being defended, it may or may not be possible to prepare an efficient defense for all possible modes of attack. If defense is not possible, then the available choices would seem to be either preemptive strike or avoidance of conflict. Defense of civilians, as stated above, is likely to be difficult to impossible. Conflict may be avoided by deterrence only in certain cases—where the opponent has a comparable amount to lose. In asymmetric situations, where opponents may have very different resources and may value them very differently, deterrence may not work at all. Conflict may also be avoided by reducing the sources of tension

Broader Context

Military activity does not take place in isolation. It is frequently motivated by non-military politics, though warlords can fight simply to improve their military position. Molecular manufacturing will be able to revolutionize economic infrastructures, creating material abundance and security that may reduce the desire for war—if it is distributed wisely.

It appears that an all-out war between molecular manufacturing powers would be highly destructive of humans and of natural resources; the objects of protection would be destroyed long before the war-fighting ability of the enemy. In contrast, a war between molecular manufacturing and a conventionally armed power would probably be rapid and decisive.

The same is true against a nuclear power that was prevented from using its nuclear weapons, either by politics or by anti-missile technologies. Even if nuclear weapons were used, the decentralization allowed by self-contained exponentially manufacturing nanofactories would allow survival, continued prosecution of the war, and rapid post-war rebuilding.

The line between policing and military action is increasingly blurred. Civilians are becoming very effective at attacking soldiers. Meanwhile, soldiers are increasingly expected to treat civilians under occupation as citizens (albeit second-class citizens) rather than enemy. At least in the US, paramilitary organizations (both governmental and commercial) are being deployed in internal civilian settings, such as the use of SWAT teams in some crime situations, and Blackwater in post-Katrina New Orleans.

Many molecular manufactured weapon systems will be useable for policing. Several factors will make the systems desirable for police activity: a wide range of weapon effects and intensities to choose from; less risk to police as telepresence is employed; maintaining parity with increasingly armed criminals; and increased deterrence due to increased information-gathering and surveillance. This means that even without military conflict, a variety of military-type systems will be not only developed, but also deployed and used.

It is tempting to think that the absence of nuclear war after six decades of nuclear weapons implies that we know how to handle insanely destructive weapons. However, a number of factors will make molecular manufacturing arms races less stable than the nuclear arms race—and it should be remembered that on several different occasions, a single fortuitous person or event has prevented a nuclear attack. Nuclear weapons are hard to design, hard to build, require easily monitored testing, do indiscriminate and lasting damage, do not rapidly become obsolete, have almost no peaceful use, and are universally abhorred. Molecular manufactured weapons will be easy to build, will in many cases allow easily concealable testing, will be relatively easy to control and deactivate, and would become obsolete very rapidly; almost every design is dual-use, and peaceful and non-lethal (police) use will be common. Nuclear weapons are easier to stockpile than to use; molecular manufactured weapons will be the opposite.

Interpenetrating arrays of multi-scale complex weapons cannot be stable for long. Sooner or later, and probably sooner, a perceived attack will be answered by an actual attack. Whether this mushrooms out of control into a full-scale conflict will depend on the programming of the weapon systems. As long as it is only inanimate hardware at stake, probing attacks and small-scale accidental attacks may be tolerated.

Given the amount of damage that a hostile power armed with molecular manufacturing products could do to the civilian sector, it seems likely that hostile actors will be tolerated only as a last resort, and even apparently non-hostile but untrustworthy actors will be highly undesirable. As mentioned above, an asymmetry in values may prevent deterrence from working. An asymmetry in force, such as between a molecular manufacturing and a pre-MM power, may tempt a preemptive strike to prevent molecular manufacturing proliferation. Likewise, a substantial but decreasing lead in military capability may lead to a preemptive strike. It is unclear whether in general a well-planned surprise attack would lead to rapid and/or inexpensive victory; this may not become clear until offensive and defensive systems are actually developed.

One stable situation appears to be that in which a single power deploys sufficient sensors and weapons to prevent any other power from developing molecular manufacturing. This would probably require substantial oppression of civilians and crippling of industrial and scientific capacity. The government in power would have near-absolute control, being threatened only by internal factors; near-absolute power, combined with an ongoing need for oppression, would likely lead to disastrous corruption.

Widespread recognition of the dangers of arms race, preemptive strike, and war might inspire widespread desire to avoid such an outcome. This would require an unprecedented degree of trust and accountability, worldwide. Current government paradigms are probably not compatible with allowing foreign powers such intimate access to their secrets; however, in the absence of this degree of openness, spying and hostile inspections will only raise tension and reduce trust. One possible solution is for governments to allow their own citizens to observe them, and then allow the information gained by such distributed and non-combative (and thus presumably more trustworthy) observation to be made available to foreign powers.

EXPLORING THE MOLECULAR WORLD

In a scenario in the last chapter, we saw Joel Gregory manipulating molecules in the virtual reality of a simulated world using video goggles, tactile gloves, and a supercomputer. The early twenty-first century should be able to do even better. Imagine, then, that today you were to take a really long nap, oversleep, and wake up decades later in a nanotechnological world.

In the twenty-first century, even more than in the twentieth, it's easy to make things work without understanding them, but to a newcomer much of the technology seems like magic, which is dissatisfying. After a few days, you want to understand what nanotechnology is, on a gut level. Back in the late twentieth century, most teaching used dry words and simple pictures, but now-for a topic like this-it's easier to explore a simulated world. And so you decide to explore a simulation of the molecular world.

Looking through the brochure, you read many tedious facts about the simulation: how accurate it is in describing sizes, forces, motions, and the like; how similar it is to working tools used by both engineering students and professionals; how you can buy one for your very own home, and so forth. It explains how you can tour the human body, see state-of-the-art nanotechnology in action, climb a bacterium, etc. For starters, you decide to take an introductory tour: simulations of real twentieth-century objects alongside quaint twentieth-century concepts of nanotechnology.

After paying a small fee and memorizing a few key phrases (any variation of "Get me out of here!" will do the most important job), you pull on a powersuit, pocket a Talking Tourguide, step into the simulation chamber, and strap the video goggles over your eyes. Looking through the goggles, you seem to be in a room with a table you know isn't really there and walls that seem too far away to fit in the simulation chamber.

But trickery with a treadmill floor makes the walk to the walls seem far enough, and when you walk back and thump the table, it feels solid because the powersuit stops your hand sharply at just the right place. You can even feel the texture of the carvings on the table leg, because the suit's gloves press against your fingertips in the right patterns as you move. The simulation isn't perfect, but it's easy to ignore the defects. On the table is (or seems to be) an old 1990s silicon computer chip.

VISION AND MOTION

You feel as though you're falling toward the chip's surface, shrinking rapidly. The world grows blurrier, then everything seems to go wrong as you approach the molecular level. First, your vision blurs to uselessness-there is light, but it becomes a featureless fog. Your skin is tickled by small impacts, then battered by what feel like hard-thrown marbles. Your arms and legs feel as though they are caught in turbulence, pulling to and fro, harder and harder. The ground hits your feet, you stumble and stick to the ground like a fly on flypaper, battered so hard that it almost hurts. You asked for realism, and only the built-in safety limits in the suit keep the simulated thermal motions of air molecules and of your own arms from beating you senseless.

"Stop!" gives you a rest from the suit's yanking and thumping, and "Standard settings!" makes the world around you become more reasonable. The simulation changes, introducing the standard cheats. Your simulated eyes are now smaller than a light wave, making focus impossible, but the goggles snap your vision into sharpness and show the atoms around you as small spheres. (Real nanomachines are as blind as you were a moment ago, and can't cheat.) You are on the surface of the 1990s computer chip, between a cell and two blocky nanocomputers. Your simulated body is 50 nanometers tall, about 1/40,000,000 your real size, and the smaller nanocomputer is twice your height.

The simulation keeps bombarding you with air molecules, but the standard settings leave out the sensation of being pelted with marbles. A moment ago you were stuck tight to the ground by molecular stickiness, but the standard settings give your muscles the effective strength of steel-at least in simulation-by making everything around you much softer and weaker.

The tourguide says that the only unreal features of the simulation have to do with you-not just your ability to see and to ignore thermal shaking and bombardment, but also your sheer existence at a size too small for anything so complex as a human being. It also explains why you can see things move, something about slowing down everything around you by a factor of 10 for every factor of 10 enlargement, and by another factor to allow for your being made stronger and hence faster. And so, with your greater strength and some adjustments to make your arms, legs, and torso less sticky, you can stand, see, feel, and take stock of the situation.

MOLECULAR TEXTURE

The ground underfoot, like everything around you, is pebbly with atom-sized bumps the size of your fingertips. Objects look like bunches of transparent grapes or fused marbles in a variety of pretty but imaginary colors. The simulation displays a view of atoms and molecules much like those used by chemists in the 1980s, but with a sharper 3-D image and a better way to move them and to feel the forces they exert. Actually, the whole simulation setup is nothing but an improved version of systems built in the late 1980s-the computer is faster, but it is calculating the same things. The video goggles are better and the whole-body powersuit is a major change, but even in the 1980s there were 3-D displays for molecules and crude devices that gave a sense of touching them. The gloves on this suit give the sensation of touching whatever the computer simulates. When you run a fingertip over the side of the smaller nanocomputer, it feels odd, hard to describe. It is as if the surface were magnetic-it pulls on your fingertip if you move close enough. But the result isn't a sharp click of contact, because the surface isn't hard like a magnet, but strangely soft.

Touching the surface is like touching a film of fog that grades smoothly into foam rubber, then hard rubber, then steel, all within the thickness of a sheet of corrugated cardboard.

Moving sideways, your fingertip feels no texture, no friction, just smooth bumps more slippery than oil, and a tendency to get pulled into hollows. Pulling free of the surface takes a firm tug. The simulation makes your atom-sized fingertips feel the same forces that an atom would. It is strange how slippery the surface is-and it can't have been lubricated, since even a single oil molecule would be a lump the size of your thumb. This slipperiness makes it obvious how nano-scale bearings can work, how the parts of molecular machines can slide smoothly.

But on top of this, there is a tingling feeling in your fingers, like the sensation of touching a working loudspeaker. When you put your ear against the wall of the nanocomputer, you flinch back: for a moment, you heard a sound like the hiss of a twentieth-century television tuned to a channel with no broadcast, with nothing but snow and static-but loud, painfully loud. All the atoms in the surface are vibrating at high frequencies, too fast to see. This is thermal vibration, and it's obvious why it's also called thermal noise.

Gas and Liquid

Individual molecules still move too quickly to see. So, to add one more cheat to the simulation, you issue the command "Whoa!", and everything around seems to slow down by a factor of ten. On the surface, you now can see thermal vibrations that had been too quick to follow. All around, air molecules become easier to watch. They whiz about as thick as raindrops in a storm, but they are the size of marbles and bounce in all directions. They're also sticky in a magnet-like way, and some are skidding around on the wall of the nanocomputer. When you grab one, it slips away. Most are like two fused spheres, but you spot one that is perfectly round-it is an argon atom, and these are fairly rare. With a firm grip on all sides to keep it from shooting away like a watermelon seed, you pinch it between your steel-strong fingers. It compresses by about 10 percent before the resistance is more than you can overcome. It springs back perfectly and instantly when you relax, then bounces free of your grip. Atoms have an unfamiliar perfection about them, resilient and unchanging, and they surround you in thick swarms.

At the base of the wall is a churning blob that can only be a droplet of water. Scooping up a handful for a closer look yields a swarm of molecules, hundreds, all tumbling and bumbling over one another, but clinging in a coherent mass. As you watch, though, one breaks free of the liquid and flies off into the freer chaos of the surrounding air: the water is evaporating. Some slide up your arm and lodge in the armpit, but eventually skitter away. Getting rid of all the water molecules takes too much scraping, so you command "Clean me!" to dry off.

NANOMATERIALS SET FOR EXPLOSIVE GROWTH

While nanoparticles are already used to make strong, lightweight plastics, their emerging applications are even more impressive. Here's how they will further enhance plastics and achieve big growth: How do you make plastics lighter and stronger? In recent years, the answer has been to imbue them with nanomaterials-extremely small structures usually from 1 to 100 nanometers, or billionths of a meter, in size. What's more, nanomaterials are not only helping produce robust, lightweight plastic composites but are also key to products as far-ranging as sunscreens and electrically conductive coatings. And these nanoscale materials are set to claim new applications in electronics, food packaging, industrial processing and other fields.

Watch for nanomaterials to figure in heat-and corrosion-resistant plastics as well as in automotive catalysts. In short, nanomaterials are on the verge of explosive growth, and you can expect plastics, among many other applications, to benefit.

Currently, nanotechnology generates $385 million in business per year in the U.S., a number that will likely swell to $3.5 billion by 2008 and $20 billion by 2013, according to a report by Consulting Resources Corp. (CRC), a Massachusetts-based market research firm. In 2008 and beyond, nanomaterials will enjoy "increasingly significant" markets in biomaterials, catalysts, diagnostics and electronics, the CRC study forecasts.

Small startups are responsible for much of the advanced work in nanomaterials. Many of these firms have developed proprietary processes to produce these diminutive particles in extremely consistent sizes, shapes and purities. Some have partnered with large global chemical and plastics suppliers, including Honeywell, DuPont, BASF, GE, Bayer, Rohm and Haas, DSM and Mitsubishi. Commercial and developmental nanoparticles can be classified into three groups-metal oxides, nanoclays and carbon nanotubes.

METAL OXIDES

Metal-oxide nanoparticles are the most highly commercialized of the three categories thus far. This group encompasses oxides of zinc, iron, cerium and zirconium, as well as mixed-metal compounds such as indium-tin oxide (ITO). These materials are currently used in polishing agents for semiconductor wafers, sunscreen formulas and scratch-resistant coatings for glass. Metal oxide developers are also eyeing new applications. For example, Degussa's recently launched internal venture aims to coat touch screens and flat panel displays with an electrically conductive nanomaterial layer using ITO nanoparticle-laden dispersions, instead of relying on the current method of sputtering, a far more costly process, says Geoffrey Varga, director of Degussa Advanced Nanomaterials, Germany. Additionally, the company is working on filling adhesive formulations with nanosized iron oxide particles to enable "bonding and debonding on command," explains Varga.

Meanwhile, a third possible application for metal oxides is in automotive catalysts. Illinois-based firm Nanophase Technologies Corp. can make narrowly distributed, nonporous, and nearly spherical nanoparticles, which would likely better withstand extremely hot, high-stress environments like catalytic converters compared to current oxide catalyst particles, which tend to be larger and less uniform. These nanoparticles could stay intact in such harsh environments, says Ed Ludwig, vice president for business development at Nanophase.

Nanoclays

The development of nanoclays is also rapidly progressing. Naturally occurring, plate-like clay particles, these materials can be loaded into plastics to make composites with enhanced strength, heat resistance, barrier properties and flame retardancy. These materials are currently used in barrier packaging for food and beverages to keep oxygen out as well as in automotive plastic composites. For example, Texas-based nanoclay company, Southern Clay Products-in collaboration with General Motors and global polyolefins producer Basell-has developed a nanocomposite thermoplastic olefin-based step assist for two GM car models.

And nanoclay makers are looking beyond automotive applications. Says Gerardo Lopez, commercial development manager at Southern Clay, "we are certainly pursuing" other uses, including flame retardants and wire and cable sheathing. Finally, a third emerging application for nanoclays is in heavy-duty industrial plastics.

Carbon Nanotubes

Carbon nanotubes are currently being used to fill electrically conductive plastics to disperse static charges in manufacturing equipment for computer disk drives and semiconductor wafers. They also eliminate the need for expensive primer coats on plastic automobile body panels, making these plastic parts conductive so that they can be electrostatically spray painted right away. While carbon nanotubes go head to head with graphite fibers in some of these applications, nanotubes require significantly lower loadings and thus better preserve polymer properties.

Nanotubes are also under development for use as flame retardants in plastics. For example, Massachusetts-based Hyperion Catalysis International is developing its multi-wall nanotubes for this particular application. "There are preliminary scientific reports that carbon nanotubes are good flame retardants in polyethylene, polypropylene and EVA (ethylene vinyl acetate)," says Patrick Collins, marketing manager for Hyperion. Additionally, the company is working to use these nanomaterials as conductive components in plastic antennas for high-speed wireless communications devices, where they could rapidly dissipate charges and allow for more efficient functioning.

Behind multi-wall carbon nanotubes in the development timeline are single-wall carbon nanotubes, which possess extraordinarily high electrical and thermal conductivity, strength, stiffness and toughness. While they are extremely challenging to mass-produce, Texas-based startup Carbon Nanotechnologies Inc. is now fabricating commercial quantities at a pilot plant. Single-wall nanotubes could potentially be used as field emitters in flat panel displays as well as in exceptionally strong industrial plastics, which could withstand high temperatures, severe chemicals, corrosive conditions, tremendous pressures and abrasion.

THE NANO FUTURE

Indeed, nanomaterials are on the verge of adding a whole new set of applications to an already impressive list that includes sunscreens, electrically conductive coatings and strong, lightweight plastic composites. With a market that many expect to explode, these nanoparticles will become essential to many more products in the near future. And you can expect plastics to improve drastically as a result.

5

Nanotechnology in Medicine

Developments in nanotechnology will result in improved medical sensors. As protein chemist Bill DeGrado notes, "Probably the first use you may see would be in diagnostics: being able to take a tiny amount of blood from somebody, just a pinprick, and diagnose for a hundred different things. Biological systems are already able to do that, and I think we should be able to design molecules or assemblies of molecules that mimic the biological system." In the longer term, though, the story of nanotechnology in medicine will be the story of extending surgical control to the molecular level. The easiest applications will be aids to the immune system, which selectively attack invaders outside tissues. More difficult applications will require that medical nanomachines mimic white blood cells by entering tissues to interact with their cells. Further applications will involve the complexities of molecular-level surgery on individual cells.

As we look at how to solve various problems, you'll notice that some that look difficult today will become easy, while others that might seem easier turn out to be more difficult. The seeming difficulty of treating disorders is always changing: Once polio was frequent and incurable, today it is easily prevented. Syphilis once caused steady physical decline leading to insanity and death; now it is cured with a shot. Athlete's foot has never been seen as a great scourge, yet it remains hard to cure. Likewise with the common cold. This pattern will continue: Deadly diseases may be easily dealt with, while minor ills remain incurable, or vice versa. As we will see, a mature nanotechnology-based medicine will be able to deal with almost any physical problem, but the order of difficulty may be surprising. Nature cares nothing for our sense of appropriateness. Horribleness and difficulty just aren't the same thing.

WORKING OUTSIDE TISSUES

One approach to nanomedicine would make use of microscopic mobile devices built using molecular-manufacturing equipment. These would resemble the ecosystem protectors and mobile cleanup machines discussed in the last chapter. Like them, they would either be biodegradable, self-collecting, or collected by something else once they were done working. Like them, they would be more difficult to develop than simple, fixed-location nanomachines, yet clearly feasible and useful. Development will start with the simpler applications, so let's begin by looking at what can be done without entering living tissues.

The skin is the body's largest organ, and its exposed position subjects it to a lot of abuse. This exposed position, though, also makes it easier to treat. Among the earlier applications of molecular manufacturing may be those popular, quasimedical products, cosmetics. A cream packed with nanomachines could do a better and more selective job of cleaning than any product can today. It could remove the right amount of dead skin, remove excess oils, add missing oils, apply the right amounts of natural moisturizing compounds, and even achieve the elusive goal of "deep pore cleaning" by actually reaching down into pores and cleaning them out. The cream could be a smart material with smooth-on, peel-off convenience. The mouth, teeth, and gums are amazingly troublesome. Today, daily dental care is an endless cycle of brushing and flossing, of losing ground to tooth decay and gum disease as slowly as possible. A mouthwash full of smart nanomachines could do all that brushing and flossing do and more, and with far less effort-making it more likely to be used.

This mouthwash would identify and destroy pathogenic bacteria while allowing the harmless flora of the mouth to flourish in a healthy ecosystem. Further, the devices would identify particles of food, plaque, or tartar, and lift them from teeth to be rinsed away. Being suspended in liquid and able to swim about, devices would be able to reach surfaces beyond reach of toothbrush bristles or the fibers of floss. As short-lifetime medical nanodevices, they could be built to last only a few minutes in the body before falling apart into materials of the sort found in foods (such as fiber). With this sort of daily dental care from an early age, tooth decay and gum disease would likely never arise. If under way, they would be greatly lessened.

Going beyond this superficial treatment would involve moving among and modifying cells. Let's consider what can be done with this treatment inside the body, but outside the body's tissues. The bloodstream carries everything from nutrients to immune-system cells, with chemical signals and infectious organisms besides.

Here, it is useful to think in terms of medical nanomachines that resemble small submarines. Each of these is large enough to carry a nanocomputer as powerful as a mid-1980s mainframe, along with a huge database (a billion bytes), a complete set of instruments for identifying biological surfaces, and tools for clobbering viruses, bacteria, and other invaders.

Immune cells, as we've seen, travel through the bloodstream checking surfaces for foreignness and-when working properly-attacking and eliminating what should not be there. These immune machines would do both more and less. With their onboard sensors and computers, they will be able to react to the same molecular signals that the immune system does, but with greater discrimination.

Before being sent into the body on their search-and-destroy mission, they could be programmed with a set of characteristics that lets them clearly distinguish their targets from everything else. The body's immune system can respond only to invading organisms that had been encountered by that individual's body. Immune machines, however, could be programmed to respond to anything that had been encountered by world medicine.

Immune machines can be designed for use in the bloodstream or the digestive tract (the mouthwash described above used these abilities in hunting down harmful bacteria). They could float and circulate, as antibiotics do, while searching for intruders to neutralize. To escape being engulfed by white blood cells making their own patrols, immune machines could display standard molecules on their surface-molecules the body knows and trusts already-like a fellow police officer wearing a familiar uniform.

When an invader is identified, it can be punctured, letting its contents spill out and ending its effectiveness. If the contents were known to be hazardous by themselves, then the immune machine could hold on to it long enough to dismantle it more completely.

How will these devices know when it's time to depart? If the physician in charge is sure the task will be finished within, say, one day, the devices prescribed could be of a type designed to fall apart after twenty-four hours. If the treatment time needed is variable, the physician could monitor progress and stop action at the appropriate time by sending a specific molecule-aspirin perhaps, or something even safer-as a signal to stop work. The inactivated devices would then be cleared out along with other waste eliminated from the body.

ELIMINATING INVADERS

Cancers are a prime example. The immune system recognizes and eliminates most potential cancers, but some get by. Physicians can recognize cancer cells by their appearance and by molecular markers, but they cannot always remove them all through surgery, and often cannot find a selective poison. Immune machines, however, will have no difficulty identifying cancer cells, and will ultimately be able to track them down and destroy them wherever they may be growing. Destroying every cancer cell will cure the cancer.

Bacteria, protozoa, worms, and other parasites have even more obvious molecular markers. Once identified, they could be destroyed, ridding the body of the disease they cause. Immune machines thus could deal with tuberculosis, strep throat, leprosy, malaria, amoebic dysentery, sleeping sickness, river blindness, hookworm, flukes, candida, valley fever, antibiotic-resistant bacteria, and even athlete's foot. All are caused by invading cells or larger organisms (such as worms). Health officials estimate that parasitic diseases, common in the Third World, affect more than one billion people. For many of these diseases, no satisfactory drug treatment exists. All can eventually be eliminated as threats to human health by a sufficiently advanced form of nanomedicine.

HERDING CELLS

Destroying invaders will be helpful, but injuries and structural problems pose other problems. Truly advanced medicine will be able to build up and restructure tissues. Here, medical nanodevices can stimulate and guide the body's own construction and repair mechanisms to restore healthy tissue.

What is Healthy Tissue?

It consists of normal cells in normal patterns in a normal matrix all organized in a normal relationship to the surrounding tissues. Surgeons today (with their huge, crude tools) can fix some problems at the tissue level. A wound disrupts the healthy relationship between two different pieces of tissue, and surgical glues and sutures can partly remedy this problem by holding the tissues in a position that promotes healing. Likewise, coronary artery bypass surgery brings about a more healthy overall configuration of tissues-one that provides working plumbing to supply blood to the heart muscle. Surgeons cut and stitch, but then they must rely on the tissue to heal its wounds as best it can.

Healing establishes healthy relationships on a finer scale. Cells must divide, grow, migrate, and fill gaps. They must reorganize to form properly connected networks of fine blood vessels. And cells must lay down materials to form the structural, intercellular matrix-collagen to provide the proper shape and toughness, or mineral grains to provide rigidity, as in bone. Often, they lay down unwanted scar tissue instead, blocking proper healing.

With enough knowledge of how these processes work (and nanoinstruments can help gather that knowledge) and with good enough software to guide the process-a more difficult challenge-medical nanomachines will be able to guide this healing process. The problem here is to guide the motion and behavior of a mob of active, living cells-a process that can be termed cell herding.

Cells respond to a host of signals from their environment: to chemicals in the surrounding fluids, to signal molecules on neighboring cells, and to mechanical forces applied to them. Cell-herding devices would use these signals to spur cell division where it is needed and to discourage it where it is not. They would nudge cells to encourage them to migrate in appropriate directions, or would simply pick them up, move them along, and deliver them where needed, encouraging them to nestle into a proper relationship with their neighbors.

Finally, they would stimulate cells to surround themselves with the proper intercellular-matrix materials. Or-like the owner of a small dog who, on a cold day, wraps the beast in a wool jacket-they would directly build the proper surrounding structures for the cell in its new location.

In this way, cooperating teams of cell-herding devices could guide the healing or restructuring of tissues, ensuring that their cells form healthy patterns and a healthy matrix and that those tissues have a healthy relationship to their surroundings. Where necessary, cells could even be adjusted internally, as we will discuss later.

Rebuilding Tissues

Again, skin provides easy examples and may be a natural place to start in practice. People often want hair where they have bare skin, and bare skin where they have hair. Cell herding machines could move or destroy hair follicle cells to eliminate an unwanted hair, or grow more of the needed cells and arrange them into a working follicle where a hair is desired. By adjusting the size of the follicle and the properties of some of the cells, hairs could be made coarser, or finer, or straighter, or curlier. All these changes would involve no pain, toxic chemicals, or stench. Cell-herding devices could move down into the living layers of skin, removing unwanted cells, stimulating the growth of new cells, narrowing unnaturally prominent blood vessels, insuring good circulation by guiding the growth of any needed normal blood vessels, and moving cells and fibers around so as to eliminate even deep wrinkles.

At the opposite end of the spectrum, cell herding will revolutionize treatment of life-threatening conditions. For example, the most common cause of heart disease is reduced or interrupted supply of blood to the heart muscle. In pumping oxygenated blood to the rest of the body, the heart diverts a portion for its own use though the coronary arteries. When these blood vessels become constricted, we speak of coronary-artery disease. When they are blocked, causing heart muscle tissue to die, we speak of someone "having a coronary," another term for heart attack.

Devices working in the bloodstream could nibble away at atherosclerotic deposits, widening the affected blood vessels. Cell herding devices could restore artery walls and artery linings to health, by ensuring that the right cells and supporting structures are in the right places. This would prevent most heart attacks. But what if a heart attack has already destroyed muscle tissue, leaving the patient with a scarred, damaged, and poorly functioning heart? Once again, cell-herding devices could accomplish repairs, working their way into the scar tissue and removing it bit by bit, replacing it with fresh muscle fiber. If need be, this new fiber can be grown by applying a series of internal molecular stimuli to selected heart muscle cells to "remind" them of the instructions for growth that they used decades earlier during embryonic development.

Cell-herding capabilities should also be able to deal with the various forms of arthritis. Where this is due to attacks from the body's own immune system, the cells producing the damaging antibodies can be identified and eliminated. Then a cell-herding system would work inside the joint where it would remove diseased tissues, calcified spurs, and so forth, then rework patterns of cells and intercellular material to form a healthy, smoothly working, and pain-free joint.

Clearly, learning to repair hearts and learning to repair joints will have some basic technologies in common, but much of the research and development will have to be devoted to specific tissues and specific circumstances. A similar process-but again, specially adapted to the circumstances at hand-could be used to strengthen and reshape bone, correcting osteoporosis.

In dentistry, this sort of process could be used to fill cavities, not with amalgam, but with natural dentin and enamel. Reversing the ravages of periodontal disease will someday be straightforward, with nanomedical devices to clean pockets, join tissues, and guide regrowth. Even missing teeth could be regrown, with enough control over cell behavior.

WORKING ON CELLS

Moving through tissues without leaving a trail of disruption will require devices able to manipulate and direct the motions of cells, and to repair them. Much remains to be learned-and will be easy to learn with nanoscale tools-but today's knowledge of cells is enough for a start on the problem of how to do surgery on cells.

Cell biology is a booming field, even today. Cells can be made to live and grow in laboratory cultures if they are placed in a liquid with suitable nutrients, oxygen, and the rest. Even with today's crude techniques, much has been learned about how cells respond to different chemicals, to different neighbors, and even to being poked and cut with needles. Conducting a rough sort of surgery on individual cells has been routine for many years in scientific laboratories.

Today, researchers can inject new DNA into cells using a tiny needle; small punctures in a cell membrane automatically reseal. But both these techniques use tools that on a cellular scale are large and clumsy-like doing surgery with an ax or a wrecking ball, instead of a scalpel. Nano-scale tools will enable medical procedures involving delicate surgery on individual cells.

ELIMINATING VIRUSES BY CELL SURGERY

Some viral diseases will respond to treatments that destroy viruses in the nose and throat, or in the bloodstream. The flu and common cold are examples. Many others would be greatly improved by this, but not eliminated. All viruses work by injecting their genes into a cell and taking over its molecular machinery, using it to produce more viruses. This is part of what makes viral illnesses so hard to treat-most of the action is performed by the body's own molecular machines, which can't be interfered with on a wholesale basis. When the immune system deals with a viral illness, it both attacks free virus particles before they enter cells, and attacks infected cells before they can churn out too many more virus particles.

Some viruses, though, insert their genes among the genes of the cell, and lay low. The cell can seem entirely normal to the immune system, for months or years, until the viral genes are triggered into action and begin the infective process anew. This pattern is responsible for the persistence of herpes infections, and for the slow, deadly progress of AIDS.

These viruses can be eliminated by molecular-level cellular surgery. The required devices could be small enough to fit entirely within the cell, if need be. Greg Fahy, who heads the Organ Cryopreservation Project at the American Red Cross's Jerome Holland Transplantation Laboratory, writes, "Calculations imply that molecular sensors, molecular computers, and molecular effectors can be combined into a device small enough to fit easily inside a single cell and powerful enough to repair molecular and structural defects (or to degrade foreign structures such as viruses and bacteria) as rapidly as they accumulate....There is no reason such systems cannot be built and function as designed."

Equally well, a cell surgery device located outside a cell could reach through the membrane with long probes. At the ends of the probes would be tools and sensors along with, perhaps, a small auxiliary computer. These would be able to reach through multiple membranes, unpackage and uncoil DNA, read it, repackage it, and recoil it, "proofreading" the DNA by comparing the sequences in one cell to the sequences of other cells.

On reading the genetic sequence spelling out the message of the AIDS virus, a molecular surgery machine could be programmed to respond like an immune machine, destroying the cell. But it would seem to make more sense simply to cut out the AIDS virus genes themselves, and reconnect the ends as they were before infection. By doing this, and killing any viruses found in the cell, the procedure would restore the cell to health.

MOLECULAR REPAIRS

Cells are made of billions of molecules, each built by molecular machines. These molecules self-assemble to form larger structures, many in dynamic patterns, perpetually disintegrating and reforming. Cell-surgery devices will be able to make molecules of sorts that may be lacking, while destroying molecules that are damaged or present in excess. They will be able not only to remove viral genes, but to repair chemical and radiation-caused damage to the cell's own genes. Advanced cell surgery devices would be able to repair cells almost regardless of their initial state of damage.

By activating and inactivating a cell's genes, they will be able to stimulate cell division and guide what types of cells are formed. This will be a great aid to cell herding and to healing tissues.

As surgeons today rely on the spontaneous, self-organizing ability of cells and tissues to join and heal the parts they manipulate, so cell-surgery devices will rely on the spontaneous self-organizing capabilities of molecules to join and "heal" the parts they put together. Healing of a surgical wound involves sweeping up dead cells, growing new cells, and a slow and genuinely painful process of tissue reorganization. In contrast, the joining of molecules is almost instantaneous and occurs on a scale far below that of the most sensitive pain receptor. "Healing" will not begin after the repair devices have done their work, as it does in conventional surgery: rather, when they complete their work, the tissue will have been healed.

HEALING BODY AND LIMB

The ability to herd cells and to perform molecular repairs and cell surgery will open new vistas for medicine. These abilities apply on a small scale, but their effects can be large scale.

Correcting Chemistry

In many diseases, the body as a whole suffers from misregulation of the signaling molecules that travel through its fluids. Many are rare: Cushing's disease, Grave's disease, Paget's disease, Addison's disease, Conn's syndrome, Prader-Labhart-Willi syndrome. Others are common: millions of older women suffer from osteoporosis, the weakening of bones that can accompany lowered estrogen levels.

Diabetes kills frequently enough to rank in the top ten causes of death in the United States; the number of individuals known to have it doubles every fifteen years. It is the leading cause of blindness in the United States, with other complications including kidney damage, cataracts, and cardiovascular damage. Today's molecular medicine tries to solve these troubles by supplying missing molecules: diabetics inject additional insulin.

While helpful, this doesn't cure the disease or eliminate all symptoms. In an era of molecular surgery, physicians could choose instead to repair the defective organ, so it can regulate its own chemicals again, and to readjust the metabolic properties of other cells in the body to match. This would be a true healing, far better than today's partial fix.

Only now are researchers making progress on another frequent problem of metabolic regulation: obesity. Once this was thought to have one simple cause (consuming excess calories) and one main result (greater roundness than favoured by today's aesthetics), but both assumptions proved wrong. Obesity is a serious medical problem, increasing the risk of diabetes mellitus, osteoarthritis, degenerative diseases of the heart, arteries, and kidneys, and shortening life expectancy. And the supposed cause, simple overeating, has been shown to be incorrect-something dieters had always suspected, as they watched thinner colleagues gorge and yet gain no weight.

The ability to lay in stores of fat was a great benefit to people once upon a time, when food supplies were irregular, nomadism and marauding bands made food storage difficult and risky, and starvation was a common cause of death. Our bodies are still adapted to that world, and regulate fat reserves accordingly. This is why dieting often has perverse effects. The body, when starved, responds by attempting to build up greater reserves of fat at its next opportunity. The main effect of exercise in weight reduction isn't to burn up calories, but to signal the body to adapt itself for efficient mobility.

Obesity therefore seems to be a matter of chemical signals within the body, signals to store fat for famine or to become lean for motion. Nanomedicine will be able to regulate these signals in the bloodstream, and to adjust how individual cells respond to them in the body. The latter would even make possible the elusive "spot reduction program" to reshape the distribution of body fat. Here, as with many potential applications of nanotechnology, the problem may be solved by other means first. Some problems, though, will almost surely require nanomedicine.

New Organs and Limbs

So far we've seen how medical nanotechnology would be used in the simpler applications outside tissues-such as in the blood-then inside tissues, and finally inside cells. Consider how these abilities will fit together for victims of automobile and motorcycle accidents. Nanomanufactured medical devices will be of dramatic value to those who have suffered massive trauma. Take the case of a patient with a crushed or severed spinal cord high in the back or in the neck. The latest research gives hope that when such patients are treated promptly after the injury, paralysis may be at least partially avoidable, sometimes. But those whose injuries weren't treated-including virtually all of today's patients-remain paralyzed.

While research continues on a variety of techniques for attempting to aid a spontaneous healing process, prospects for reversing this sort of damage using conventional medicine remain bleak. With the techniques discussed above, it will become possible to remove scar tissue and to guide cell growth so as to produce healthy arrangements of the cells on a microscopic scale. With the right molecular-scale poking and prodding of the cell nucleus, even nerve cells of the sorts found in the brain and spinal cord can be induced to divide. Where nerve cells have been destroyed, there need be no shortage of replacements. These technologies will eventually enable medicine to heal damaged spinal cords, reversing paralysis. The ability to guide cell growth and division and to direct the organization of tissues will be sufficient to regrow entire organs and limbs, not merely to repair what has been damaged. This will enable medicine to restore physical health despite the most grievous injuries.

If this seems hard to believe, recall that medical advances have shocked the world before now. To those in the past, the idea of cutting people open with knives painlessly would have seemed miraculous, but surgical anesthesia is now routine. Likewise with bacterial infections and antibiotics, with the eradication of smallpox, and the vaccine for polio: Each tamed a deadly terror, and each is now half-forgotten history. Our gut sense of what seems likely has little to do with what can and cannot be done by medical technology. It has more to do with our habitual fears, including the fear of vain hopes. Yet what amazes one generation seems obvious and even boring to the next. The first baby born after each breakthrough grows up wondering what all the excitement was about. Besides, nano-scale medicine won't be a cure-all. Consider a fifty-year-old mentally retarded man, with a mind like a two-year-old's, or a woman with a brain tumor that has spread to the point that her personality has changed: How could they be "healed"? No healing of tissues could replace a missed lifetime of adult experience, nor can it replace lost information from a severely damaged brain. The best physicians could do would be to bring the patients to some physically healthy condition. One can wish for more, but sometimes it won't be possible.

First Aid

Throughout the centuries, medicine has been constrained to maintain functioning tissues, since once tissues stop functioning, they can't heal themselves. With molecular surgery to carry out the healing directly, medical priorities change drastically-function is no longer absolutely necessary. In fact, a physician able to use molecular surgery would prefer to operate on nonfunctioning, structurally stable tissue than on tissue that has been allowed to continue malfunctioning until its structure was lost.

Brain tumors are an example: They destroy the brain's structure, and with it the patient's skills, memories, and personality. Physicians in the future should be able to immediately interrupt this process, to stop the functioning of the brain to stabilize the patient for treatment.

Techniques available today can stop tissue function while preserving tissue structure. Greg Fahy, in his work on organ preservation at the American Red Cross, is developing a technique for vitrifying animal kidneys-making them into a low-temperature, crystal-free glass-with the goal of maintaining their structure such that, when brought back to room temperature, they can be transplanted. Some kidneys have been cooled to – 30 °C, warmed back up, and then functioned after transplantation. A variety of other procedures can also stabilize tissues on a long-term basis. These procedures enable many cells-but not whole tissues-to survive and recover without help; advanced molecular repair and cell surgery will presumably tip the balance, enabling cells, tissues, and organs to recover and heal. When applied to stabilizing a whole patient, such a condition can be called biostasis. A patient in biostasis can be kept there indefinitely until the required medical help arrives. So in the future, the question "Can this patient be restored to health?" will be answered "Yes, if the patient's brain is intact, and with it the patient's mind."

Sandra Lee Adamson of the National Space Society has her eyes on distant goals. Some have proposed that travel to the stars would take generations, preventing anyone on Earth from ever making the trip. But she notes that biostasis will "give hope to some fearless adventurers who will risk suspension and subsequent reanimation so they can see the stars for themselves."

PLAGUE INSURANCE

Medical nanotechnologies promise to extend healthy life, but if history is any guide, they may also avert sudden massive death. The word plague is rarely heard today, except in relation to AIDS; it calls up visions of the Black Death of the Middle Ages, when one third of Europe died in 1346-50. A virulent influenza struck in 1918, half lost in the news of the First World War: how many of us realize that it killed at least 20 million? People often act as though plagues were gone for good, as if sanitation and antibiotics had vanquished them. But as doctors are forever telling their patients, antibiotics kill bacteria, but are useless for viruses. The flu, the common cold, herpes, and AIDS-none has a really effective treatment, because all are caused by viruses. In some African countries, as much as 10 percent of the population is estimated to be infected with the AIDS-causing HIV virus. Without a cure soon, the steep rise in deaths from AIDS still lies in the future. AIDS stands as a grim reminder that the great plagues of history are not behind us.

The Threat

New diseases continue to appear today as they have throughout history. Today's population, far larger than that of any previous century, provides a huge, fertile territory for their spread. Today's transportation systems can spread viruses from continent to continent in a single day. When ships sailed or churned their way across the seas, an infected passenger was likely to show full-blown disease before arrival, permitting quarantine. But few diseases can be guaranteed to show themselves in the hours of a single aircraft flight.

So far as is known, every species of organism, from bacterium to whale, is afflicted with viruses. Animal viruses sometimes "jump the species gap" to infect other animals, or people. Most scientists believe that the ancestors of the AIDS virus could, until recently, infect only certain African monkeys. Then these viruses made the interspecies jump. A similar jump occurred in the 1960s when scientists in West Germany, working with cells from monkeys in Uganda, suddenly fell ill. Dozens were infected, and several died of a disease that caused both blood clots and bleeding, caused by what is now named the Marburg virus. What if the Marburg virus had spread with a sneeze, like influenza or the common cold? We think of human plagues as a health problem, but when they hit our fellow species, we tend to see them from an environmental perspective. In the late 1980s, over half the harbor-seal population in large parts of the North Sea suddenly died, leading many at first to blame pollution. The cause, though, appears to be a distemper virus that made the jump from dogs. Biologists worry that the virus could infect seal species around the world, since distemper virus can spread by aerosols-that is, by coughing-and seals live in close physical contact. So far its mortality rate has been 60 to 70 percent.

What of AIDS itself: Could it change and give rise to a form able to spread, say, as colds do? Nobel Laureate Howard M. Temin has said, "I think that we can very confidently say that this can't happen." Nobel Laureate Joshua Lederberg, president of Rockefeller University in New York City, replied, "I don't share your confidence about what can and cannot happen." He points out that "there is no reason a great plague could not happen again....We live in evolutionary competition with microbes-bacteria and viruses. There is no guarantee that we will be the survivors."

6

Nanostructures

A nanostructure is an object of intermediate size between molecular and microscopic (micrometer-sized) structures. In describing nanostructures it is necessary to differentiate between the number of dimensions on the nanoscale.

Nanotextured surfaces have one dimension on the nanoscale, i.e., only the thickness of the surface of an object is between 0.1 and 100 nm. Nanotubes have two dimensions on the nanoscale, i.e., the diameter of the tube is between 0.1 and 100 nm; its length could be much greater. Finally, spherical nanoparticles have three dimensions on the nanoscale, i.e., the particle is between 0.1 and 100 nm in each spatial dimension.

The terms nanoparticles and ultrafine particles (UFP) often are used synonymously although UFP can reach into the micrometre range. The term 'nanostructure' is often used when referring to magnetic technology.

The nanostructures considered in this book are magnetic and characterized by *structural* length scales ranging from a few interatomic distances to about one micrometer. The basic length unit is the *nanometer* (1 nm = 10–3μm = 10–9m), corresponding to about four interatomic Fe-Fe distances. Magnetic nanostructures pose experimental challenges, exhibit interesting physical phenomena, and have many present or potential applications.

An important aspect is that structural lengths affect, but only partly determine, the magnetic length scales encountered in the structures. Examples are domains in semihard nanoparticles, where both the domain size and the domain-wall thickness may be smaller than the particle size, and polycrystalline soft-magnetic nanostructures, where the magnetic correlation length is much larger than the crystallite size.

Due to rapid progress in the fabrication and processing of nano structures, it is now possible to realize a broad variety of geometries, crystalline textures, and chemistries. For a given geometry, the structures can be fabricated using a variety of magnetic materials (compounds), with different local magnetic properties and crystalline textures. The Appendix presents some magnetic materials of interest in nanomagnetism.

Basic Definitions and Units

The magnetic moment m of the atoms in a nanostructure nearly exclusively originates from the electrons in the partially filled inner shells of transition or rare-earth metals. There are both spin (*S*) and orbital (*L*) contributions, but since L is much smaller than S in most iron-series transition-metal magnets, the magnetic moment is often equated with the spin polarization. The situation is similar to that encountered in bulk magnets, although both *S* and *L* may be modified at surfaces and interfaces.

As in infinite solids, nuclear moments are much smaller than electron moments and can be ignored safely for most applications. The magnetic moment of an atom is created by intra-atomic or Hund's-rules exchange, which favors parallel spin alignment on an atomic scale. In addition, ferromagnetism requires *interatomic* exchange, to ensure parallel alignment of the moments of neighboring atoms.

Figure: ***Nanostructures***

The resulting net moment gives rise to the magnetization M = J/ìo which is defined as the moment per unit volume. In bulk ferromagnets, the competition between interatomic ex change and thermal disorder leads to the vanishing of the spontaneous mag netization *MS* = |M(r)| at a well-defined sharp Curie temperature *TC*. How ever, the existence of a Curie-point singularity is limited to infinite bodies, and in nanostructures, the concepts of magnetic phase transitions must be reevaluated.

Spin-orbit coupling in combination with local crystal-field interactions gives rise to magnetic anisotropy. In the simplest case, magnetic anisotropy is of the uniaxial type and described by the lowest-order anisotropy constant *K*1. This constant is equal to the energy density necessary to turn the mag netization from the easy to a hard magnetization axis. In addition, there are shape-anisotropy contributions of magnetostatic origin. In nanostructures, surface, inter-face, and shape anisotropy contributions are often important, particularly in cubic materials, where the lowest-order bulk anisotropy is zero.

An applied magnetic field H changes the magnetization by rotating the local moment. Since the magnetic anisotropy yields energy barriers between different magnetization states or spin configurations M(r), the field dependence of the net magnetization exhibits *hysteresis.* Important hysteretic properties are remanence *M*r, defined as the zero-field magnetization after saturation in a strong magnetic field, and the coercivity *H*c. The latter is defined as the reverse magnetic field at which the volume-averaged magnetization of an initially saturated magnet reaches zero. Some other hysteretic properties of specific nanostructuring offers many tools to tune hysteretic properties. For example, the coercivity of advanced magnetic nanostructures varies from about 1 ìT to several T.

Analytical and numerical aspects of the hysteresis of magnetic nano-structures. The structural length scales of nanomagnets are intermediate between interatomic and macroscopic distances, but nanomagnetism cannot be reduced to a mixture of atomic-scale and macroscopic phenomena. For example, most extrinsic properties are realized on a nanoscale, and nano structuring is used to produce optimized hard, soft, information-storage and sensor materials. A dynamic aspect of this interplay between atomic (or intrinsic) and hysteretic (or extrinsic) phenomena is that equilibration times vary from less than 1 ns to millions of years.

This determines, for example, the magnetic switching time of spin-electronics structures and the lifetime of information stored in magnetic recording media. This is related to the thermal instability of the magnetization direction known as superparamagnetism (Bean and Livingston 1959). This effect occurs in very small particles and is strongly temperature dependent.

In this book, the preferred *length unit* is 1 nm (10–9m), but 1 Å = 0.1 and 1 ìm = 1000 nm are also used, particularly for submicron features having sizes of several 100 nm. Recording densities are also measured in bytes per square inch or bytes per square centimeter (1 in –2 = 6.452 cm –2). Both SI and Gaussian units will be used for *magnetic quantities,* with explicit conversions occasionally included in square brackets. This also includes the parallel or alternate use of the SI unit tesla for coercivity and magnetization, as compared to the correct but cumbersome unit A/m. The latter corresponds to B = ìo(M + H), and the former is obtained by incor porating ìo into M and H, so that one actually considers ìo M and ìo H. In the Gaussian system, multiplying the magnetization (emu/cm3) by the dimensionless number 4 changes the unit to gauss. (This is similar to measuring the perimeter of an island in miles and its diameter in feet.) Some everyday numerical *conversion rules* are: (i) 1 T = 10 kOe H" 0.8 MA/m (coercivity), (ii) 1 emu/cm3 = 1 kA/m and 1 T = 10 kG H" 0.8 MA/m (magnetization), (iii) 1 T 2= 100 MGOe H" 800 kJ/m3(energy product).

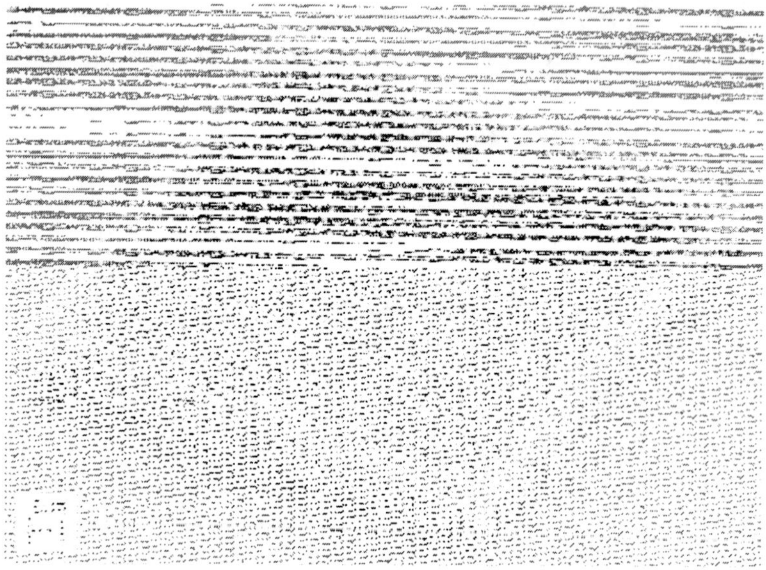

Figure: ***Silicon nanostructures***

Finally, an SI susceptibility of 1 corresponds to a Gaussian susceptibility of 4. In the above con versions, we have used the numerical relation that 10/4 = 0.7958 H" 0.8.

Magnetic Nanostructures

Magnetic nanostructures can be produced in a wide range of geometries. Several chapters of this book deal with the fabrication, investigation, and application of individual geometries, such as nanowires and patterned thin films. The following paragraphs briefly characterize typical geometries, mention some systems of practical or scientific interest, and provide links to the individual chapters of this book.

Nanoparticles, Clusters, and Molecular Magnets

Small magnetic particles exist in nature or are produced artificially. Nanoparticles have sizes ranging from a few nanometers to submicron dimensions, whereas molecular magnets contain a few magnetic atoms in well-defined atomic environments. Clusters are inter mediate structures, with less well-defined atomic environments but ex hibiting atomic features such as facets. Examples of naturally occurring nanoparticles are magnetite (Fe_3O_4) nanoparticles precipitated in bacteria, insects and higher animals, and magnetite and other oxide particles responsible for rock magnetism.

Nanobio-magnetics is concerned not only with questions such as the role of magnetite particles for horizontal and vertical orientation of animals but also with important medical issues, such as local drug administration and cancer diagnosis. The small remanent magnetization of magnetic rocks, first analyzed by Neel in the 1940's, is exploited, for example, in archeo-magnetic dating and to monitor changes in the Earth's magnetic field. Small oxide particles, less than 10 nm in diameter, are observed in gels having the nominal composition $FeO(OH){\cdot}nH_2O$. Fine particles are also encountered in meteorites.

Some artificially produced magnetic nanoparticle structures are Fe in Al_2O3 and so-called "elongated single-domain particles." Interesting applications of small particles are stable colloidal suspensions known *ferrofluids. A* variety of materials can be used, such as Fe_3O4, $BaFe_{12}O_{19}$, Fe, Co, and Ni, and a typical particle size is 10 nm. Most ferrofluids are based on hydrocarbons or other organic liquids, whereas water-based ferrofluids are more difficult to produce. They are used as liquids in bearings and to monitor magnetic fields and domain configurations.

Granular Nanostructures

Embedded clusters, granular materials, and other bulk nanostructures are of great importance in nanoscience. The structural correlation lengths of typical nanocomposite materials range from about 1 nm in x-ray amorphous structures to several 100 nanometers in submicron structures. Magnetic glasses and atomic-scale defect structures are beyond the scope of nanomagnetics, but they are of indirect interest as limiting cases and because nanomagnetic phenomena have their quantum-mechanical origin in atomic-scale magnetism.

Depending on grain size and microchemistry, granular nanostructures are used for example as permanent magnets (Nd-Fe-B), soft magnets (Fe-Cu-Nb-Si-B), and magnetoresistive materials (Co-Ag). There are two types of exchange-coupled permanent magnets: isotropic magnets, which exhibit random anisotropy and remanence enhancement, and oriented hard-soft composites, which utilize exchange coupling of a soft phase with a high magnetization to a hard skeleton. Closely related systems with many potential applications are magnetic clusters deposited in a matrix. For example, the narrow size distribution of 10–20% makes this material interesting as a granular media for magnetic recording. A well-known *soft-magnetic* nanocomposite is the

"Yoshizawa" alloy Fe73.5Si13.5B9Cu1Nb3, which consists of DO3-structured FeSi grains embedded in an amorphous matrix.

Particle Arrays and Functional Components

Two-dimensional arrays of nanoparticles are of interest as scientific model systems and have many present or future applications. For example, advanced magnetic recording media can be characterized as a complex array of magnetic particles, and interest in dot arrays has been sparked by the search for ever-increasing storage densities in magnetic recording. In very small dots, quantum-mechanical effects are no longer negligible, and there are phenomena such as quantum-well states. These effects are of interest in quantum computing and spin electronics.

Most easily produced and investigated are submicron dots made from iron-series transition metals, such as Ni, but it is also possible to use metallic alloys, such as Permalloy, and to reduce the dot size to less than 100 nm. The dots may form square or hexagonal arrays, or structures such as corrals. Among the investigated phenomena are the properties of individual dots and interdot interactions. A related class of nanostructures are *antidots,* that is, holes in a film rather than dots on a film.

Potential applications include magnetic recording, sensors, magnetic and quantum computing, micron- and submicron-size mechanical devices, short-wavelength optics, and spin electronics. Other "functional" building blocks are, for example, nanojunctions, spin valves, and tips for magnetic-force microscopy (MFM tips).

Nanowires

There is a smooth transition from elongated dots and thin-film patches to nanowires. Magnetic nanowires have present or potential applications in many areas of advanced nanotechnology, including patterned magnetic media, magnetic devices, and materials for microwave technology. There a various methods to produce nanowires, such as deposition on vicinal surfaces and electrodeposition, including electrodeposition into porous alumina templates.

Much of the early work on magnetic nanowire arrays was concerned with exploratory issues, such as establishing an easy axis for typical preparation conditions, the essential involvement of shape anisotropy, as opposed to magnetocrystalline anisotropy, and the description of magnetostatic interactions between wires. More recently, attention has shifted towards the understanding of magnetization processes, such as the transition from curling-type to quasi-coherent nucleation, the influence of deposition-dependent polycrystallinity of typical transition-metal nanowires. Some other interesting phenomena are magnetic-mode localization, as evident *e.g.* from experimental activation volumes, spin-waves, and current-induced magnetization reversal.

FINITE ELEMENT MICROMAGNETICS

The finite element method requires the discretization of the spatial domain &! with "finite elements" – a regular triangulation. For two dimensional problems triangles and rectangles, in three dimensions tetrahedral and hexahedral elements are commonly used. Also a mixture of different types of elements

is possible, but after the evaluation of various other implementations and for simplicity a tetrahedral discretization has been implemented. As compared to finite difference methods the finite element mesh may be entirely unstructured, which makes the modeling of complicated geometries and irregular microstructures more convenient. However, the creation of the geometrical model and its triangulation are still very demanding tasks, which require sophisticated (commercial) tools. Patran by MSC.Software and GID (there is a free academic version available) are suitable software packages for which import filters have been implemented. The required resolution of the finite element mesh – the maximum size of the finite elements – is determined by the smallest features which might occur in the solution of the PDE. These features have to be properly resolved and the approximation of the exact solution by the test functions has to be sufficiently accurate to give meaningful results.

In micromagnetics these limits are defined by three typical lengths:

*l*bex="*A*/*K*1, δBW= π*l*bex(1)δBW is the width of Bloch type magnetic domain walls (typically found in bulk hard magnets), *A* is the exchange constant and *K*1 is the magnetocrystalline anisotropy constant.

*l*nex="2μ0*A*/*J*2s, δNW= π*l*nex(2)δNW is the width of Neel walls (typically found in soft magnetic materials and thin films), *J*s is the saturation polarization. *l*thex="*AJ*s *H*th, *H*th="2α*k*B*T*"*t*γ*J*s*l*3(3)*l*thex is the thermal exchange length with the thermal field *H*th, which depends on the Gilbert damping constant α, the temperature *T*, time step of the numerical time integration scheme "*t*, the gyromagnetic ratio γ, and the spatial correlation length *l* (which equals the cell size). In the most general case, the maximum cell size of the finite element mesh has to be smaller than the minimum of the three lengths defined above. However, if the structure of the solution is roughly known, it is possible to use an "adapted" mesh for the simulations. It has a high resolution (small elements) in areas with very small features (e.g. domain walls) and a low resolution (large elements) in other areas where the solution (magnetization) is very uniform. As a result, the number of nodes (and therefore the number of unknowns) and, thus, the time required for the simulation can be greatly reduced. For dynamic time dependent problems a rigid adapted mesh is often not suitable, because the solution changes over time and the smallest features of the solution move through the mesh. If they leave the high resolution mesh and reach areas with larger finite elements, artificial pinning on the mesh will occur and give wrong results. Thus, an adaptive mesh refinement method is required, which changes the structure of the mesh during the simulation and adapts the resolution of the mesh to the solution. It can be shown, that these methods lead to (almost) optimal complexity and give most accurate results with the smallest numerical (computational) effort. Adaptive mesh refinement methods are still a very active research area, and they have been successfully applied also in numerical micromagnetics. However, this adaptive meshing technique has not been implemented (yet) in our micromagnetics package.

FIRST-PRINCIPLE STUDIES OF NANOSTRUCTURES

Recent advances in theoretical methods and high-performance computing allow for reliable first principle calculations of complex nanostructures. Nanostructured materials are characterized by a fascinating diversity of geometries, but here we restrict ourselves mainly to first-principle cal culations for nanoparticles and clusters, nanowires and nanocontacts. Nanoscale multilayers are also discussed very briefly, although multilayers are often considered as a subfield of thin-film physics rather than nanoscience. We also ignore nanotubes, because most of the work in this direction has been done on non-magnetic carbon nanotubes.

Nanoparticles and Clusters

In magnetic recording, a bit of information corresponds to a small area or volume of ferromagnetic material with a well-defined and as high as possible magnetic moment. Furthermore, below a given size the grains tend to become superparamagnetic at room temperature, and the stored information would become unstable. It is known, both experimentally and theoretically, that the magnetic moments and the magnetic anisotropy energy per atom are often enhanced in small metallic clusters, as compared to the bulk crystalline state. Nanosize transition-metal particles, between the molecular and metallic states, are of particular interest in this context due to their intrinsically high magnetic moment. The properties of these clusters depend strongly on its surrounding. We will compare magnetic properties of free-standing clusters, clusters supported on substrate, and clusters embedded in the media.

Studies of small free-standing FeN, CoN and NiN clusters have revealed that their magnetic moments are significantly larger than the corresponding bulk magnetizations. Non-vanishing magnetic moments have even been observed in small clusters of some 4d transition metals, for example Pd, Ru and Rh, which are non-magnetic in the solid state. For example, Kumar *et al.* performed ab initio calculations on the growth and magnetic behavior of Pd clusters having up to 147 atoms and found an icosahedral growth and a ferromagnetic behavior in these clusters.

The magnetization of free clusters of elemental metals exhibits an oscillatory behavior with increasing cluster size. The origin of these oscillations is the successive addition of atomic shells with increasing cluster size. The variation of the magnetic moment in the outermost shell is due to the changing coordination number of surface atoms. As a rule, surface atoms with lower coordination (lower number of nearest neighbors) have larger moments, and the corresponding theory of magic numbers seems to explain experimental observations.

Using a tight-binding Hubbard Hamiltonian in the unrestricted Hartree-Fock approximation, Pastor *et al.* have shown the size and structural dependence of magnetic properties of free Crn, Fen, and Nin ($n < 15$) clusters, and also the exchange interaction and local environments effects in Fen clusters. The finite-temperature magnetism of small clusters, which is remarkably different from that of bulk systems, has also been studied in terms of a similar approach by taking into account both electronic and structural excitations. A great advantage of the TB methods is that they easily can be combined with molecular dynamics calculations, enabling the investigation of relaxation effects which proved to be important in determining the magnetic moments and the MAE of transition metal clusters. As mentioned above, the anisotropy of the free standing clusters is very complicated. Experimental studies predict a very complicated switching behavior, suggesting a strong competition between different contributions to the anisotropy energy.

The main reason is that different pairs of spin-orbit split levels yield anisotropy contributions of opposite sign, so that the net anisotropy is the sum of many positive and negative contributions. Phenomenological studies complement this point and yield good agreement with experiment. The supercell-based studies of supported clusters show that the magnetic moment tends to increase as well. For example, small Co nanoclusters may exhibit a large orbital moment and large anisotropy on Pt surface.

The pros and cons of large orbital moments in applications of nanostructures was discussed by Andriotis *et al.*. Magnetic-nanoparticle patterned media are extensively studied as a potential breakthrough technology for high-density storage of information. The key requirement is that the individual nanoparticles exhibit both a large magnetic moment and a high anisotropy. More generally, in the last decades, theoretical and experimental studies have shown that the reduction of size and dimensionality can give rise to a large range of novel materials. This has opened the possibility of generating specifically designed nanomaterials with tailored properties. Understanding the conditions for the development of such properties is crucial in order to be able to properly adjust and control these parameters, in particular for magnetic applications.

The embedding technique based on the Korringa-Kohn-Rostoker (KKR) Green-function method in the local spin-density approximation (LSDA) has been applied to the magnetism of transition metal adatoms and clusters deposited on surfaces. The main feature of this approach is that the interaction between adatoms and host surface atoms can be analyzed within first-principles electronic structure calculations.

An accurate cal culation of the total energy in terms of full potential or full charge density schemes made possible the investigations of the energetics of adatoms. As compared to TB methods, an obvious drawback of the embedded KKR technique is that, with respect to computational limitations, the number of the atoms in the cluster is restricted to about less than 100. Furthermore, the inclusion of structural relaxations is exceedingly difficult. On the level of a fully relativistic spin-polarized electron theory, recently, strongly enhanced orbital magnetism and MAE of adatoms and small clusters on Ag and Au (100) surfaces have been reported.

Spin polarized electronic structure calculations of the Co clusters embedded in Cu matrix have been reported by Qiang *et al.*. The authors found that the magnetization of the embedded Co clusters in Cu matrix is somewhat lower than the magnetization of the bulk Co, *i.e.* the effect is opposite to the free cluster case. This is caused by the slight suppression of the moment of cobalt sites in the interface region. The magnetization oscillations are similar to the free case in theory if shell by shell filling of the clusters is expected. However, this condition is not satisfied in case of embedded clusters and magic number theory does not work.

Moreover, there is always some interdiffusion on the interface between atoms of the matrix and atoms of the cluster in case of larger clusters. Frequently it creates a magnetically dead layer at the interface. The exchange interactions between metallic clusters in metallic media exhibit oscillatory dependence on the intercluster distance like it happens in Co clusters embedded in the Cu matrix. The calculated inter-cluster exchange interactions shows both ferromagnetic and aniferromagnetic exchange between clusters, and the values of exchange at short distances is of the order of interatomic exchange in the bulk cobalt. The spin-glass type behavior in this and similar systems arises from frustration.

This is very likely responsible for the lowering of the magnetization with the increase of cobalt concentration from 10 to 50%. Nanostructured permanent magnets use the idea of mixing soft magnetic materials and hard magnetic materials in a composite structure. The goal is to combine a large magnetocrystalline anisotropy of hard magnetic materials and large magnetization of the soft magnetic materials in order to obtain large energy product (*BH*)max, which requires large magnetization and large coercivity. This idea was first proposed by Kneller and Hawig, by estimating the effect in multilayers. Micromagnetic calculations based on parameters obtained from first principles show that FePt/Fe nanostructures may exhibit a (*BH*)max as high as 90 MGOe. These changes are experimentally observed in various systems, and Section 2.3 in the following chapter discusses some systems.

The properties of embedded nanoclusters can be controlled by the properties of the matrix. For example, the matrix can exert uniaxial or biaxial strain on the embedded nanoclusters. The corresponding strain can modify the properties of nanostructure considerably. For example, Zeng *et al.* show that applied stress can modify the Curie temperature and anisotropy of FePt.

Nanowires

The progress in experimental techniques has made it possible to synthesize stable ultrathin metal nanowires with diameter down to atomic sizes and of high aspect ratios. An example is single-crystalline silver nano wires having a width of 0.4 nm width and deposited into the pores of an organic template. Kondo *et al.* have observed novel helical multishell structures in the suspended ultrathin gold and platinum nanowires. This is a relatively mature field of studies with thousands of articles published in the last few years, and there are various review articles available.

While magnetic properties of free-standing nanoclusters have been extensively studied experimentally, studies of magnetic nanowires are largely limited to theoretical calculations. One exception are wires freed from templates, where mechanical stability excludes very thin diameters. Decades ago, Weinert and Freeman investigated the electronic structures and magnetism of the linear Fe and Ni chains by using the self-consistent full-potential linearized augmented plane wave (FLAPW) method. They predicted large magnetic moments of 3.3 ìB and 1.1 ìB for the Fe and the Ni chains, respectively. Free-standing Coatomic chains show an en hanced magnetization of 2.3 ìB as well.

It is interesting that first-principle calculations predict ferromagnetism in monatomic nanowires of the nonmagnetic transition metals Ru, Rh, and Pd, with respective mean-field moments of 1.1, 0.3, and 0.7 ìB at the equilibrium bond length. An analysis of the band structures indicates that the nanocontact superparamagnetic state suggested by calculations should affect the ballistic conductance between tips made of Ru, Rh or Pd, leading to possible temperature and magnetic-field dependent conductance. One method to produce nanowires is the growth on the steps of the crystalline surface cut with the high Miller index.

Spisak and Hafner investigated the position dependent magnetic moments of the Fe nanowire on a vicinal Cu(117) by using the pseudopotential method. From the total energy comparison, they found that the system of the Fe nanowire placed on the inner comer site of the Cu(117) step is most stable with a magnetic moment of 2.80 ìB /atom, while the system of the Fe nanowire placed on the top of the terrace (formed by (001) plane) of Cu(l 17) is unsta ble with a magnetic moment of 2.96 ìB/atom.

Jin *et al.* investigated the electronic structures and magnetism of an Fe nanowire along the direction on a Cu(00l) surface (Fe chain) by using the FLAPW method and found large magnetic moments of the Fe atoms:

3.11 and 3.39 ìB /atom for the Fe [010] chain and the free-standing nanowires (free-Fe[010] chain), respectively. Since free-standing nanowires of elemental metals are one-dimensional objects, they exhibit magnetic properties similar to free standing clusters, *i.e.* its magnetic moment increases especially for the atoms at surface. The lower coordinated atoms have higher moment, while atoms inside of the nanostructure have magnetic moment close to the bulk values. Recently, investigations of the electronic structures and magnetism of an Fe chain along the direction on a Cu(001) were performed by using the FLAPW method within the generalized gradient approximation. Hong *et al.* have calculated properties of Co atomic wires on the Cu(001) surface. They show that magnetic moment of atomic size wires can be affected by the substrate. The magnetocrystalline anisotropy of Co wires supported on Cu substrate has opposite sign from the free standing case, i.e. the magnetization is along the wire on the Cu surface and perpendicular in free-standing case. Gambardella *et al.* investigated a quasi-one-dimensional Co chain at the

Pt(111) step edge. The symmetry breaking at the step leads to an easy magnetization axis at an odd angle of about 20 degrees towards the Pt step. The orbital and spin moment may have different orientation in this system. This reflects the specific character of the crystal-field interaction in these systems.

It is difficult to tackle the problem of embedded nanowires. An example is Co nanowires fully coated with atomic Cu shells. Wang *et al.* find that Cu atoms occupy the surface, while Co atoms prefer to occupy the interior of the nanowires to form the perfect-coated structures. The coated Cu atomic layer leads to a large variation of the magnetic moment of the Co nanowires, depending on the structure and the thickness of Cu layers.

A single row of Co atoms in the center of nanowire is found to be nonmagnetic when coated with two Cu layers, while all other investigated Co nanowires remain magnetic, albeit with reduced magnetic moments. The transport properties of nanowires are of technological importance and have attracted significant attention in the recent years. Band structure gives simple solution to the analysis of the ballistic transport of periodic nanowires because the number of the bands crossing the Fermi surface is equal the number of quantum of conductance. However, the situation in nanocontacts is more complicated.

Nanocontacts

The understanding of spin transport in order to use it in electronics devices has been a field of intense research for the past few years. Interesting effects were reported in electrodeposited Ni nanocontacts, where resistance changes by a huge value on application of a small magnetic field at room temperature. This has followed after effort was focused on the magnetoresistance effect of electrodeposited nanocontacts. Smaller magnetoresistance ratios, but interesting from the point of view of applications, were reported also for Co, Fe and Fe_3O_4 nanocontacts.

There have been several and partially exclusive attempts to explain the experimental data. One possibility considered is the scattering of the electrons in a domain wall. The width of a domain wall in a nanocontact is predicted to be of the order of the nanocontact size and such a reduction of the domain wall width is likely to enhance the magnetoresistance. However, this effect is very small, unless the domain walls contain atomically sharp feature, especially layers with reduced exchange. Magnetostriction effects could also lead to an increased magnetoresistance, but most probably cannot explain the experiments. Alternatively, a thin domain wall pinned in a magnetic dead layer together with the possibility of 100% polarization of the Ni d electrons was also proposed as a possible explanation. Using a free-electron model and a linear response theory Zhuravlev *et al.* investigated a spin-dependent electronic transport in magnetic nanocontacts in the ballistic regime of conduction.

In particular, they show that the presence of a nonmagnetic region in the nanocontact separating two ferromagnetic electrodes can lead to a spin blockade resulting in very large values of magnetoresistance. Other possible explanations are a large magnetostriction in the experimental samples, or the presence of the defects in the constriction. The latter may result in the resonant tunneling and provide large MR.

In order to describe the nanocontacts theoretically, several methods have been developed by various groups. Recently, Burton *et al.* investigated the electrical resistance of a constrained DW in a nanocontact using micromagnetic modeling and ballistic conductance calculations.

They found that for electrodes with significantly different magnetic anisotropy, the DW can be trapped by the nanoconstriction (domain-wall pinning). They showed that the DW width and, therefore, the conductance can be controlled by an applied magnetic field. The first-principle approach is based on the tight-binding (TB) formulation. An important conclusion coming from TB models is that conductance of the single-atomic contacts is proportional to the number of valence orbitals available at the Fermi energy. *Ab-initio* density functional (DF) calculations with jellium electrodes have been proposed by Lang *et al.* and Kobayashi *et al.*, which addressed single-atom contacts, atomic chains of Al, Na, and C.

The formation mechanisms of atomic chains made from different types of elements such as Ni, Pd, Pt, Cu, Ag and Au were studied by means of molecular dynamic simulations. Also, Mehrez *et al.* and Brandbyge *et al.* presented fully self-consistent DF calculations of the conductance of atomic contacts treating the electronic structure of the whole system (electrodes and the constriction region) on the same footing. The ballistic conductance through Ni, Co, and Fe nanocontacts within a semiempirical tight-binding model proposed by Velev and Butler shows that the interplay of the contact and the domain-wall resistance can produce very large giant magnetoresistance ratios.

However, the very giant magnetoresistance is limited to geometries in which the nanocontacts are very narrow and have very small aspect ratios, similar to the predictions in. Solanki *et al.* calculated the electronic structure and conduc tance of atomic-size Ni contacts using a real-space tight-binding LMTO method and recursion technique within the frame work of density functional theory. They used the Landauer-Buttiker approach to calculate the conductance. Their results of spin-dependent conductance as a function of energy show ballistic bulk-like behavior. The appreciable dependence of the conductance on the structure of nanocontact is reported in their work. Similar conclusion was made by Emberly *et al.* considering carbon nanowires. We should notice that there are several research groups which could not reproduce giant MR in the nanocontacts experimentally.

Multilayers, Surface, and Interfaces

Magnetic thin films and multilayers have attracted much interest in the context of nanomagnetism, although they are often considered as a separate branch of condensed matter physics. Ultrathin magnetic films, epitaxially grown on nonmagnetic substrates, are prototype systems for inves tigating magnetism in two dimensions, and the possibility to produce films and multilayers with perpendicular magnetic anisotropy for magnetooptic recording applications is of technological importance. The topic of perpendicular magnetic anisotropy is very challenging, since the magnetostatic dipolar interactions always prefer in-plane orientation of the easy axis of magnetization.

Since the pioneering experimental work by Gradmann and the theoretical predictions of Gay and Richter much effort has been spent investigating Fe, Co, and Ni films grown on noble-metals substrates. Among the hetero-epitaxial systems that show strong per pendicular magnetic anisotropy (PMA) are Co/Au(111) thin films and superlattices, Ni/Cu(001) and Ni/Cu(111) thin films, Co/Au(111) thin films and multilayers, bcc Fe/Ag(001) and fcc Fe/Cu(001) thin films. Sabirianov and Jaswal carried out first-principles studies of exchange-coupled hard-soft multilayers predicting very large energy products needed for hard-magnet applications.

The discovery of the giant-magnetoresistance (GMR) effect has further renewed interest in the theory of transport in layered and granular systems. Several approaches to transport in layered systems are currently being used. The phenomenon of GMR is promising for its applications, such as in nonvolatile magnetic random-access memories in the information technology industry or as reading heads and various kinds of sensors in the recording and car industries, respectively.

The GMR effect has been observed mostly in the diffusive transport regime in which the mean free path is much smaller than the dimension of the so-called active part of the multilayer system. In the ballistic regime, the mean free path is larger than the dimension of the active part of the multilayer system. Transport in multilayers has been studied extensively and review on the subject covers the topic thoroughly.

The magnetocrystalline anisotropy of multilayered systems has been also calculated by many groups. Újfalussy *et al.* performed fully relativistic spin-polarized local spin density calculations for Co monolayer on Au(111). They obtained an enhancement of perpendicular magnetic anisotropy as a function of the Au coverage. They showed the close relationship between the anisotropies of orbital magnetic moments and the anisotropy energies, thus interpreting their results in terms of familiar perturbation theory.

Lorentz and Hafner presented calculations of the magnetic structure and of the uniaxial and planar anisotropies of thin films of Fe on Cu(001) substrates. They have used the real space recursion calculations using TB-LMTO to calculate the anisotropy. Their results for Fe monolayers reveal the importance of the surface and interface anisotropies, including the effect of nonmagnetic coverages. Essentially, this is a crystal-field effect, similar to that in bulk compounds consisting of atomic layers with different magnetic properties.

Stepanyuk and Hergert have performed molecular dynamics simulations of Co nanostructures on Cu(001) and Cu(111) surfaces. They first performed *ab-initio* calculations of the electronic structure in the framework of density functional theory and then used the results for dynamical simulations.

EXCHANGE ENERGY

Since we know the external field explicitly, we can just simply add it to the other contributions to the effective field.

Demagnetizing Field and Magnetostatic Energy

The demagnetizing field is a little more complicated to handle, because it is an "open boundary problem" with one of its boundary conditions at infinity. The magnetostatic potential can be calculated directly by integration using the Green's function, but this method is very slow and expensive in storage. Fast multipole methods have been developed to overcome these problems. Fredkin and Koehler proposed a hybrid finite element/boundary element method that requires no finite elements outside the magnetic domain Ω. A comparison of these methods can be found in.

Since the matrices for the exchange and anisotropy energies depend only on the (time independent) material parameters and the geometry, they need to be calculated only once at the beginning. In order to save time and memory, they can also be assembled into a single combined matrix, if the energies and fields are not required separately.

METHODS OF ELECTRONIC-STRUCTURE CALCULATIONS

Electronic structure methods for studies of nanostructures can be divided broadly into supercell methods and real-space methods. Supercell methods use standard k-space electronic structure techniques separating periodically repeated nanostructures by distances large enough to neglect their inter actions. Direct space methods do not need to use periodic boundary conditions. Various electronic structure methods are developed and applied using both approaches. In this section we will shortly discuss few popular but powerful electronic structure methods: the pseudopotential method, linear muffintin orbital and related methods, and tight-binding methods.

Linear Muffin-Tin Orbital and Related Methods

Basis-set methods may be divided into two categories, depending on whether they (i) use wave function expansions in some set of fixed basis functions like atomic orbitals, Gaussian and plane waves, or (ii) expand the wave function in a set of energy and potential dependent partial waves as done in Korringa-Kohn-Rostoker (KKR) and the augmented plane wave (APW) methods. The KKR and APW methods require computational efforts which, despite attempts to improve the efficiency, are barely feasible and limited to sp-like valence and conduction electrons. Computationally, this can be remedied by the addition of localized orbitals to the plane wave basis set. But such a hybrid scheme is neither elegant nor in accordance with the chemical and physical intuition based on the smooth trends observed through the periodic table. It is necessary to use the self-consistent methods for computing one electron eigenval-ues and eigenvectors with speed and accuracy. The so-called linear methods of band theory satisfy the requirements rather well. This is true for the linear-muffin-tin-orbital method (LMTO). This method is linearized version of the KKR method. Almost identical with the solid state LMTO method is the augmented spherical wave (ASW) method of Williams *et al.*.

The LMTO has also been extended to treating crystalline impurities with the Green's function technique by Koeing *et al.;* and Gunnarson *et al.* have used it for both metal and semiconductor hosts. Harris, Casula and Herman and Springborg *et al.* have developed the LMTO method for clusters and molecules. For surfaces and thin films, LMTO techniques have been devised by Fuziwara and Femando *et al.*. It was also discovered that the conventional solid-state LMTO basis set can be transformed exactly into orthogonal, tight binding (TB) and minimal basis sets, which simplifies and generalizes the solid-state LMTO method considerably.

The use of the atomic-sphere approximation (ASA), where the single-electron potential is modeled as a superposition of spherical potentials inside overlapping space-filling spheres, makes LMTO methods com putationally very efficient. Where this approximation is applicable, the LMTO-ASA method is presumably the most efficient procedure available for solving the density functional equations to a reasonably high degree of accuracy. However, a full-potential treatment going beyond the ASA is needed for many systems, including surfaces and impurities, and for total-energy changes associated with phenomena such as phonon distortions and atomic relaxations etc.

A number of full potential methods have been developed, namely, FPLAPW, FPLMTO and FPLO. A real-space version of the full-potential approach was implemented by Beck.

Plane-Wave Pseudopotential Methods

The pseudopotential method relies on the separation (in both energy and space) of electrons into core and valence electrons and implies that most physical and chemical properties of materials are determined by valence electrons in the interstitial region. One can therefore combine the full ionic potential with that of the core electrons to give an effective potential (called the pseudopotential), which acts on the valence electrons only. On top of this, one can also remove the rapid oscillations of the valence wavefunctions inside the core region such that the resulting wavefunction and potential are smooth.

Beyond a chosen cutoff radius, the all-electron and pseudofunctions (potential and wavefunction) are identical, while inside the core region both the pseudopotential and pseudowavefunction are smoothly varying. After the construction of these pseudofunctions for a single atom and ensuring that their scattering properties are almost identical to those of the all-electron functions, they can be used in any chemical environment.

The pseudopotential method has various advantages. Eliminating the core electrons from the problem reduces the number of particles that must be considered in the Kohn-Sham (KS) equations for the effective one-electron potential. For example, a pseudopotential calculation for bulk silicon (with 10 core and 4 valence electrons) requires the calculation of 4 occupied bands at each k-point, while an all-electron approach would require the calculation of 14 occupied bands. More importantly, the smooth spatial variation of the pseudopotential and pseudowavefunction allows the use of computationally convenient and unbiased basis, such as plane wave basis sets or grids in space.

The plane-wave pseudopotential method has proven to be an excellent computational tool for solving large scale electronic structure problems. Notable strengths of the method are the ability to use fast Fourier transform to update the Kohn-Sham equations, the lack of dependence on the basis on atom positions, and the clear control of convergence with the cutoff energy determined by the shortest wavelength mode. However, the method encounters difficulties in treating widely varying length scales. This issue is especially relevant for surfaces, clusters, and the hard pseudopotentials of first row elements or transition metals, which vary rapidly near the nucleus.

A real-space version of the pseudopotential method was developed by Chelikowsky.

TIGHT-BINDING METHODS

The traditional approach to obtain the electronic structure of a periodic solid with *ab initio* pseudopotentials has been to solve the Kohn-Sham (KS) equations in momentum space using a plane-wave basis set. There are many advantages of plane-wave basis sets. For example, due to their completeness and orthonormality, they allow convenient and unbiased representation of the charge density and wavefunctions for calculating operator matrix elements.

For non-periodic and localized systems, such as defects, surfaces, clusters, quantum dots, using a plane wave basis set requires an artificial periodicity in order to implement standard electronic structure algorithms. This artificial periodicity is introduced through the use of large unit cells called *supercells.* While the supercell approach works well for localized systems, it is typically necessary to consider a very large supercell. This results in a plane-wave basis replicating not only the relevant electronic states but also vacuum regions imposed by the supercell. A much more efficient method to implement for investigating the electronic structures of localized systems is to use real space methods such as the recursion methods and the moments methods. These methods do not require symmetry and their cost grows linearly with the number of inequivalent atoms being considered. For these reasons, real space methods are very useful for a description of the electronic properties of complex systems, for which the usual k-space methods are either inapplicable or extremely costly.

Real-space methods are generally impractical, but they are extremely efficient when the system in consideration is well described by a tight-binding (TB) Hamiltonian. For a long time, parameterized TB Hamiltonians in conjunction with real space methods have been used to study the properties of complex transition metal alloys. In some cases, simple model TB Hamiltonians are very useful to stress the essential physics of the phe nomenon in question and even to give the correct trends found in real systems. But in other cases, the effects are too subtle and the results obtained by the use of these parameterized Hamiltonians can be misleading. Even when trends can be well described, to obtain quantitative results a more exact Hamiltonians is usually needed. This has been a major problem regarding real-space electronic structure calculations. Usually, the involved parameters needed to build the TB Hamiltonian, such as hopping integrals, are obtained from a linear combination of atomic orbitals (LCAO) fit to more exact k-space calculations or are adjusted to fit experimental results.

When one considers complex systems, for which no exact results are available, parameters are usually obtained from a simple similar system and then transferred to the more complex system in question using the Slater-Koster rules and some other empirical formulas which are known to be roughly obeyed. The results were often encouraging, but the lack of a solid theoretical background to justify the procedure left some fundamental questions unanswered. For example, how could one simulate the crystal-field splitting which depends on local environment and can not be transferred from one calculation to another, and how do we treat the wave function?

The solution to these problems is not unique and several interesting points of view have developed. However, major progress towards obtaining a TB-Hamiltonian based on a solid theoretical understanding of the problem came in 1984 with the advent of the LMTO-ASA-TB formalism. This gave rise to a new era regarding the quantitative description of the electronic structure of complex systems by using real-space methods.

MAGNETIC PROPERTIES

In this chapter we describe how basic or "intrinsic" magnetic properties are calculated from first principles. Technological applications, for example in magnetic recording, tend to require a large magnetization, thermal stability, and control over the coercivity. The reversal process is described usually by micromagnetic modeling which takes into account dipole-dipole inter actions. The description of magnetization reversal is beyond the scope of our paper.

Magnetic properties can be understood and fully described from microscopic considerations using *ab-initio* electronic-structure methods. The local magnetic moment in combination with intra-atomic exchange defines the magnetization, interatomic exchange describes the thermal stability, and the coercivity reflects the magnetic anisotropy of the system. These interactions have different energy scales. The intra-atomic exchange is of the order of 2 eV, interatomic exchange of the order of 10 meV, and anisotropy of the order of 1 meV.

However, all of them are extremely important to describe the behavior of a magnet. The description of nanomagnets requires new approaches. First, nanostructures are not periodic and tend to have large surface-to-volume ratios. Because of this the magnetization is not uniform across the nanostructure, local magnetic moments differ from site to site, exchange coupling varies throughout the nanostructure, and the anisotropy can be quite different from bulk or surface anisotropies. Second, it is hard to define properties in the similar fashion as in the bulk or as in case of molecules.

Magnetic Moment

The magnetic moment is given by the difference between the filling of majority and minority spin bands, which is directly related to the ferromag netic exchange splitting between the bands. There is also an orbital-moment contribution to the moment, but in 3d-based magnets, it is often small due to quenching. Total energy minimization over all occupied band states yields crystal structure and magnetic ordering. Magnetic coupling and magnetic moment are sensitive functions of the interatomic spacing and the atomic volume, respectively. Both are dominated by the Coulomb and by exchange interactions at small distances. The moment is determined pri marily by intraatomic exchange interactions and the Curie temperature by the interatomic exchange interactions.

The most common approach to calculate the magnetization from first principles is to use the local density formalism, where the spin dependent part of the total energy is approximated by an expression derived for the electron gas. The corresponding exchange-correlation potential depends only on the majority and minority spin densities. Different magnetic struc tures, such as paramagnetic, ferromagnetic and various antiferromagnetic configurations, exhibit different total energies, and the lowest energy configuration obtained by local density theory represents the observed magnetic state in most cases.

Room-temperature ferromagnetism occurs in rather few elements, Fe, Co, Ni, Gd, and several other rare earths are ferromagnetic at low temperatures. However, many alloys are ferromagnetic.

The stability of the ferromagnetism of itinerant magnets can be explained by the *Stoner criterion,* which takes the DOS at the Fermi level, $D(EF)$, and an atomic exchange integral I as input. The transition from para-magnetism to ferromagnetism is favourable for $D(EF)\ I > 1$. Then the system can lower its energy by bringing enough majority spin electrons down in energy by opening up the ferromagnetic exchange splitting.

The Stoner criterion explains why Fe, Co and Ni are singled out for ferromagnetism. Several other elements are close to fulfilling the criterion, for example Pd. Since the DOS is different for different atomic sites, nanostructures may exhibit different local moments and complicated magnetic transitions.

With increasing atomic volume, one approaches the free atom limit where Hund's first rule postulates maximum spin, so that the individual spins of the electrons in a shell are aligned parallel.

More generally, Pauli's exclusion principle implies that electrons with parallel spins have different spatial wavefunctions, reduces the Coulomb repulsion and is seen as exchange interaction. When the atoms are squeezed into a solid, some of the electrons are forced into common spatial wavefunctions, with antiparallel spins and reduction of the overall magnetic moment. At surfaces and interfaces, the reduced coordination reverses this effect, and a part of the atomic moment is recovered.

Anisotropy

Magnetic anisotropy is one of the most important properties of a magnetic material. It is the energy necessary to deflect the magnetic moment in a single crystal from the easy to a hard direction. The main contribution to the anisotropy is *magnetocrystalline anisotropy.* The easy and hard directions arise from the interaction of the spin magnetic moment with the crystal lattice that is due to spin-orbit coupling.

In crystals, the magnetocrystalline anisotropy energy is given by a series expansion in terms of the angles between the direction of magnetization and the crystal axes. It is often sufficient to represent the anisotropy energy in an arbitrary direction by just the first two terms in the series expansion. These two terms each have an empirical constant associated with them called the first- and second order uniaxial anisotropy constants K1 and K2 respectively. In cubic crystals, the lowest-order term is the fourth-order term, also known as the "cubic" K1.

There is also some shape anisotropy due to magnetostatic dipole interactions. This anisotropy is important in some nanomagnets, for example in elongated nanoparticles, but unrelated to the electronic structure. A contribution caused by spin-orbit coupling and closely related to magnetocrystalline anisotropy is *magnetoelastic anisotropy.* Mechanical stress creates a strain which amounts to a lattice distortion and yields a correction to the magnetocrystalline anisotropy. Surface anisotropy is a manifestation of magnetocrystalline anisotropy, too.

Starting with Brooks (1940), calculations of the magnetocrystalline anisotropy energy (MAE) of itinerant magnets have developed in two different approaches have been developed since the pioneering studies of Takayama, Bohnen, and Fulde and of Gay and Richter. The first approach uses perturbation theory within a semi-empirical tight-binding framework, justified by the argument that the MAE is small compared to a characteristic band-width. The second approach relies on *ab initio* spin-polarized total-energy calculations, with spin-orbit coupling included either self-consistently within the scalar-relativistic approximation or as a final perturbation to a calculation neglecting spin-orbit coupling and using the force theorem.

The main difficulty with this approach arises from the fact that a change in the direction of the magnetic moment changes the occupation of the eigenstates only in a narrow region of the Brillouin zone, hence the *k*-space integrations for the total energy are only very slowly convergent. To avoid the necessity to use a huge number of *k*-points, Wang, Wu, and Freeman introduced a state tracking procedure, using information on the change of the band structure with increasing spin-orbit interaction to extrapolate the Brillouin-zone integrals.

The description of the magnetic anisotropy of nanoparticles is more complicated than that of bulk systems, partially because shape and surface anisotropies are of the same order of magnitude as the magnetocrystalline anisotropy. The ratio of atoms on the surface to the number of atoms in the core of a nanoparticle is very large.

This changes the electronic structure of the core atoms as well and its anisotropy may be quite different from the bulk systems. Surface anisotropy may cause drastic change in magnetization and reverse properties of nanostructures. Even in a naive magnetostatic approach, surface anisotropy causes the renormalization of switching parameters.

SPIN STRUCTURE AND NONCOLLINEARITY

The term spin structure refers to the ferromagnetic, antiferromagnetic or ferri-magnetic alignment of the atomic spins. In ferromagnets, all spins are parallel and the atomic moments add, whereas ferromagnets and antiferromagnets are characterized by two or more sublattices with opposite moments. This leads to a reduction or absence of a net magnetic moment. Competing exchange interactions in periodic crystals and in disordered magnets give rise to noncol-linear spin arrangements. Examples are helimagnetic order in perfect crystals, which is caused by the competing interactions between next and more distant neighbors, and spin-glass behaviour in magnets with atomic scale disorder.

Deviations from parallel or antiparallel spin alignment may also occur at surfaces and interfaces. Further, it is important to distinguish between zero temperature magnetism and finite temperature magnetic order. It is difficult to calculate finite-temperature magnetic properties from first principles, particularly Curie and Neel temperatures. In the atomic limit, for example using the Heisenberg model, one assumes a magnetic moment on each atom that persists beyond Curie temperature. In the paramagnetic phase, the magnetization vanishes by orienting the moments at random.

The other extreme, that is the Stoner model of band like ferromagnetism, assumes that the magnetic moment gradually decreases at elevated tem perature and vanishes at the Curie temperature. It yields Curie temperatures that are several times higher than observed. The basic drawback is the assumption that the magnetic order disappears by spin flip excitations across the Stoner gap. They cost more energy than is actually needed to disorder magnetic moments.

More realistic models contain small domains with local order whose orientations are starting to fluctuate when approaching the Curie temperature. Some degree of local correlation remains even above TC. Such a picture brings in spin wave excitations (magnons) which are not contained in the ground state energy bands. The efforts to model the spin structure just below and above the Curie point from first principles give qualitatively correct Curie temperatures, even if the method used in the paper is more appropriate to very weak itinerant ferromagnets such as $ZrZn^2$.

Interatomic Exchange

Magnetic order occurs because the magnetic moments of neighbouring atoms couple by interatomic exchange. As the intra-atomic exchange responsible for the magnetic moment, it reflects the overlap of electrons in combination with Pauli's exclusion principle. Depending on the positive or negative sign of the exchange, the interaction between the ions will force the individual moments into parallel (ferromagnetic) or antiparallel (antiferromagnetic) alignment.

The direct exchange between overlapping orbitals was proposed first. Later other types of interatomic exchange were identified such as an indirect or RKKY-type exchange, superexchange, double exchange, anisotropic exchange, and higher-order exchange (such as biquadratic). All of them may be present in a system and responsible for a variety of effects governing the magnetic structure and finite-temperature behaviour of magnet. Direct exchange operates between moments which are close enough to have sufficient overlap of their wavefunctions. it gives a strong but short range coupling which decreases rapidly as the ions are separated.

Direct exchange as originally envisaged by Heisenberg, is always ferromagnetic, but there is a negative hopping contribution. When the atoms are very close together, the electrons minimize the single-electron interaction by spending a considerable time between the nuclei. Since the electrons are then at the same place in space at the same time, Pauli's exclusion principle requires them to possess opposite spins. This gives rise to antiparallel alignment, corresponding and therefore negative or antiferromagnetic exchange. Note that the direct intra-atomic exchange between electrons in the same atom J is positive, leading to Hund's first rule. Indirect exchange couples moments over relatively large distances. It is the dominant exchange interaction in metals where there is little or no direct overlap between neighbouring magnetic electrons. It therefore acts through an intermediary which in metals is the conduction electrons (itinerant electrons). This type of exchange is also known as the Ruderman, Kittel, Kasuya and Yoshida (RKKY) interaction, especially when mediated by free electrons.

This spin polarization in the itinerant electrons is felt by the moments of neighboring magnetic ions within range, leading to an indirect coupling. Since RKKYexchange oscillates from positive to negative as the separation of the ion changes, the magnetic coupling can be ferromagnetic or antiferromagnetic. In rare-earth metals whose magnetic electrons in the 4f shell are shielded by the 5s and 5p electrons, direct exchange is rather weak and insignificant and indirect exchange via the conduction electrons gives rise to magnetic order in these materials.

Electronic Structure of Nanomagnets

Recent advantages in the preparation and characterization of thin films, surfaces and nanostructures have lead to an intriguing question: is it possible to fabricate 'designer solids' by controlling materials on the atomic scale, that is layer by layer, row by row, and ultimately atom by atom? Engineered molecules are common place in biochemistry, and the same idea can be apply to solids and electronic materials. Electronic properties of semi conductor devices have been controlled by heterostructures, quantum wells and superlattices.

Magnetism as a cooperative phenomenon also lends itself to manipulation in small structures, where neighboring atoms can be replaced systematically by species with stronger or weaker magnetism. In fact, a class of magnetic/ non-magnetic multilayers termed "spin valves" has been introduced into magnetic storage devices. The wave functions of electrons undergo changes on confining them to dimensions comparable with their wavelength. In semiconductors, confinement and quantization phenomena are visible already at dimensions greater than 200 nm, whereas in metals they typically are seen at 1 nm. In fact, the Fermi wavelength of typical metals has atomic dimensions, but beat frequencies with the lattice can be an order of magnitude larger.

Consider the formation of low dimensional electronic states by quantization. Confining electrons to small structures causes the continuous bulk bands to split up into discrete levels, for example quantum well states in a slab. For N atomic layers in the slab there are N levels. In order to exhibit two dimensional behavior there should be only a single level within about 6kB T of the Fermi level.

Several levels within the Fermi cutoff would already approach a three dimensional continuum. For a coarse estimate of the corresponding slab thickness, one may set the energy E of the lowest level equal to kB T. For room temperature (E = kB T = 0.026 eV), one obtains a de Broglie wavelength $h/p = h/(2mE)$ ½ 8 nm, which is comparable with the spatial extent of the lowest quantum state. Thus, both the high electron density in magnetic metals and the requirement of room temperature operation for quantum devices point to dimensions of a few nanometers. Note, however, that Fermions are quantum-mechanical objects at any temperature below EF/kB.

The contribution of first-principle calculations to physics, chemistry, materials science, and, recently, geology and biology is more important than ever. The main reasons are the steady increase in computer power and the continuous progress in methodology (both efficiency and accuracy of algorithms and approximations). However, the application of these methods to investigate structural, electronic, and optical properties of nanostructures has not been straightforward due to extensive computational needs and to the new physics inherent in the nanometer and subnanometer region.

One particularly useful extension to overcome the computational demand imposed by localized nanostructures has been the introduction of methods based on a real-space implementation, such as the higher-order finite-difference pseudopotential methods. In the most favorable cases, quantum mechanical formulations of the electronic structure of atomic systems scale as the cube of the number of atoms of the system. This makes it very difficult to reach system sizes larger than a few hundreds of atoms, and is therefore a huge barrier for the study of problems in nanoscale materials.

The way out of this trap was found several years ago, when a number of ideas suggested the possibility of developing approximate although accurate schemes to reduce the computational cost to linear scaling. These so called *O*(*N*) methods have matured since those first proposals, and now constitute a viable route for studying systems with unprecedented size.

MAGNETIC THIN FILMS AND MULTILAYERS

Magnetic thin films and multilayers can be classified as magnetic nano-structures, too, but it is common to treat homogeneous thin films and multilayers as a separate branch of magnetism, intermediate between nano-magnetism and surface magnetism. However, many recently developed and investigated nano-structures are thin-film nanostructures. Examples are self-assembled thin-film nanostructures, patterned nanomagnetic thin films, hard-magnetic thin-film nanostructures and thin films for magnetic recording.

Semihard thin films are used in magnetic recording media and have, more recently, attracted attention as tools for magnetic information pro cessing. In addition, on a length scale of a few interatomic distances, there is a variety of interesting thin-film effects, such as vicinal and interface anisotropies, moment modifications at surfaces and interfaces, thickness-dependent domain-wall and coercive phenomena, interlayer exchange-coupling, and finite-temperature magnetic ordering. A specific example is the nanoscale exchange-coupling or "exchange-spring" effects in multilayers.

Fabrication and Characterization

The broad variety of magnetic nanostructures corresponds to a diverse range of processing methods. The suitability of individual methods depends on the length scale and geometry of the nanostructures. In addition, each method is usually restricted to a relatively narrow class of magnetic materials. *Granular nanostructures* are produced by methods such as mechanical alloying and chemical reactions. A traditional though somewhat cumber some method to fabricate *nanoscale particle arrays* of magnetic, dots, and wires on thin films is nanolithography. Other examples are molecular-beam epitaxy, the use of STMs, and chemical vapor deposition. The call for well-characterized large-area arrays of nanoparticles has stimulated the search for advanced production methods such as laser-interference lithography (LIL), where laser-intensity maxima effect a local transformation of a nonferromagnetic material into ferromagnetic islands. Another development is the use of focused ion-beam milling (FIB) to create small particles and particle arrays with well-de-fined properties.

Thin-film *nanowires* are comparatively easily obtained by depositing magnetic materials on vicinal surfaces and by exploiting structural anisotropies of the substrate. They can be produced with thicknesses down to one or two monolayers. Electrodeposition of magnetic materials into porous alumina may be used to produce regular wire arrays (see Sellmyer *et al.* 2001 and references therein). Other ways of fabricating cylindrical nanowires include deposition into molecular sieves, track-etched polymer mem branes, and mica templates.

By electrodeposition into porous anodic alumina it is now possible to produce hexagonal Fe, Co, and Ni nanowire arrays with diameters ranging from 4 to 200 nm, and lengths of up to about 1 ìm, and variable center-to-center spacings of the order of 50 nm. The resulting materials are of interest as magnetic recording media, for optical and microwave applications, and as electroluminescent display devices. Aside from the above-mentioned iron-series transition-metal elements, there is interest in depositing alloys and multilayers, such as Fe/Pt, into porous templates.

The structural and magnetic characterization of magnetic nanostructures is the main focus of Ch. 5 and of various sections and subsections through out the book. Structural correlation lengths can be probed for example by X-ray diffraction, small-angle neutron scattering (SANS) and electron microscopy. Magnetic measurements are performed with the methods known from bulk and surface magnetism, although some techniques must be adjusted to the small signals from certain structures. Examples are vibrating sample magnetometry (VSM), magneto-optical Kerr effect (MOKE) measurements, and SQUID magnetometry. Some methods, such as magnetic-force magnetometry, are nanospecific and presently being applied to measure hysteresis loops of nanoscale magnetic particles.

Applications

Magnetic nanostructures are used in the form of traditional magnetic materials, such as hard and soft magnets, and in specific functional structures, such as sensors. Hard or *permanent magnets* are used, for example, in electromotors, hard-disk drives, loudspeakers, windshield wipers, locks, refrigerator magnets, and microphones. Some applications, such as toys, do not usually require high-performance magnets, but hard-disk drives and other high-tech applications require highly sophisticated rare-earth permanent magnets with well-defined nanostructures. Compared to the highly anisotropic hard magnets, *soft magnets* exhibit very low magnetic anisotropy. They are widely used for flux guidance in permanent-magnet and other systems, in transformer cores, and for high-frequency and microwave applications, and in recording heads. In advanced soft-magnetic materials, nanostructuring is used to reduce magnetic losses by controlling anisotropy, eddy-current losses, and other properties. A key application and driving force of magnetic nanotechnology is *magnetic recording media.* They are used not only for audio-visual technology, for example in audio and video tapes, but also in computer technology, for example in hard-disks.

Aremarkable increase in areal density of many orders of magnitude in the last two decades has relied heavily on nanostructuring of media and read and write heads. Artificial nanostructuring is a way of creating completely new technologies. One area is spin electronics, and various types of nanostructures, such as multilayers and nanojunctions, are being used or investigated in this context. The magnetoresistance of metallic thin films, granular nanostructures and magnetic oxides are exploited in sensors, and a problem of current interest is spin injection into nonferromagnetic metals and magnetic semiconductors. Other recent developments are magnetic nanostructures for quantum computing, multiferroics (where nanoscale effects are exploited to synergize electric and magnetic degrees of freedom), and nanoparticle ferrofluids for cancer treatment, guided by a magnet and delivering high local doses of drugs or radiation. Nanoscale effects are also exploited in micro-electromechanical systems (MEMS) and magnetic-force nanotips made from CoPt.

THE MAGIC OF MAGNETIC MULTILAYERS

In recent years there has been a renaissance in the study of magnetic multilayers—primarily because of the discoveries of enhanced or giant magnetoresistance and oscillatory interlayer exchange coupling in magnetic multilayered structures consisting of alternating thin ferromagnetic and nonferromagnetic metallic layers. These discoveries have led to intense theoretical and experimental studies of an extraordinary range of structures over the past few years.

Giant magnetoresistance (GMR) is displayed by a wide variety of inhomogeneous magnetic nanostructures consisting of magnetic entities, for example, thin layers or particles, separated by thin nonferromagnetic metallic spacer layers. The original observation of GMR was made in single-crystalline (100) Fe/Cr "sandwich" and superlattice structures grown by molecular beam epitaxy (MBE) in ultrahigh-vacuum deposition chambers.

It had previously been found that the magnetic moments of the Fe layers in a (100)-oriented Fe/Cr/Fe sandwich are aligned antiparallel to one another in zero field when the Cr layer is about 9 Å thick. When a sufficiently large magnetic field is applied, the moments of the Fe layers become aligned parallel with one another. The resistance of such sandwiches or superlattices comprising many such sandwiches depends upon the magnetic arrangement of the magnetic layers and was observed to be higher when the moments are aligned antiparallel to one another. The magnetoresistance of such structures is much larger than that of the intrinsic magnetoresistance of the Fe layers themselves and is especially large at low temperatures.

While the observation initially appeared to be somewhat esoteric, it was found soon afterward that similar results could be obtained in polycrystalline Fe/Cr structures grown by the much simpler technique of magnetron sputtering. These experiments also revealed that GMR could be observed in a wide variety of transition-metal magnetic multilayers. In addition, the experiments showed, in contrast to the earlier work on MBE-grown structures, that the magnitude of the GMR effect oscillated as the thickness of the nonferromagnetic spacer layers between the ferromagnetic layers was increased.

This oscillation was shown to be caused by an oscillation in the sign of the interlayer exchange coupling between the ferromagnetic layers. The coupling was shown to oscillate between antiferromagnetic and ferromagnetic coupling such that the magnetic moments of successive ferromagnetic layers were either parallel (ferromagnetic) or antiparallel (antiferromagnetic) in small magnetic fields. Oscillatory coupling was shown to be a very general property of almost all transition-metal magnetic multilayered systems in which the nonferromagnetic layer comprises one of the *3d, 4d,* or *5d* transition metals or one of the noble metals. The oscillation period was found to be just a few atomic layers, typically about 10 Å, but varying up to ~20 Å. Only those multilayers for which the interlayer coupling is antiferromagnetic display significant giant magnetoresistance effects. It is only in those systems that the relative orientation of the magnetic moments of neighboring layers is significantly altered by applying a magnetic field.

Magnetron sputtering allowed the study of many different magnetic multilayered systems. One of these systems, namely ferromagnetic cobalt layers separated by thin copper layers, was found to exhibit very large GMR effects even at room temperature. Values of GMR in Co/Cu multilayers exceed 110% at room temperature[2]. While the largest GMR values require magnetic fields exceeding ~20 kOe, magnetoresistance values of 50-60% are obtained in fields of several hundred oersteds and values of ~20% or more in fields of a few tens of oersteds.

These lower values are obtained by using thicker Cu layers, for which the interlayer exchange coupling is weaker. Such large magnetoresistance values at room temperature make such multilayers attractive candidates for a variety of technological applications. Moreover, the fact that these multilayers are readily formed using deposition techniques compatible with large-scale manufacturing makes them even more alluring.

Magnetoresistive materials are used in a wide variety of applications for detecting magnetic fields. For some applications, such as the detection of the rotation of a spinning object, as for example in anti-lock brake systems in automobiles, relatively large magnetic fields can be used.

For other applications, such as the detection of magnetic bits in a hard disk drive, materials whose resistance is sensitive to quite small magnetic fields are needed. Conventional materials such as Ni-Fe alloys exhibit resistance changes at room temperature of just a few percent in magnetic fields of a few oersteds. GMR sandwiches can achieve sensitivities to such fields of perhaps as much as five times greater than conventional materials.

GMR multilayered structures have already found their way into leading-edge hard disk drive products. Early in November 1997, IBM was the first company to announce GMR magnetic recording read heads in a family of disk drive products, designated Deskstar* 16GP. First customer shipment began in January 1998. In these products, the use of GMR read heads allows the reading of extremely small magnetic bits at an areal density of 2.69 gigabits per square inch.

The capacity of a hard disk drive is largely determined by the areal density in conjunction with the size and number of disks or platters within the drive. GMR read heads allow more than 3.2 gigabytes of data to be stored on each 95-mm-diameter disk of the Deskstar 16GP disk drives.

The resulting data storage capacity of these "3.5-inch form-factor" drives is 16.8 gigabytes. The areal density of hard disk drives, driven by the rapid evolution of the various technologies involved, is increasing at a compound growth rate of about 60% per year, and by the year 2001 is expected to reach 10 gigabits per square inch.

This issue of the *IBM Journal of Research and Development* discusses some of the work carried out over the past few years, largely within IBM, concerning GMR, oscillatory interlayer exchange coupling in magnetic multilayers, and related effects.

The first paper, by Allenspach and Weber, discusses some experiments which probe the origin of the oscillatory interlayer coupling in single-crystal (100)-oriented Co/Cu/Co sandwiches prepared on single-crystal Cu substrates. These experiments have shown that the strength and period of the oscillatory coupling could largely be explained from the well-known electronic properties of Cu. In these experiments, the thickness of the Cu and Co layers could be varied by fractions of an atomic layer, and it was observed that the magnetic properties of this system depend on such minute variations.

The second paper, by Jones, discusses some of the different theoretical approaches used to understand oscillatory interlayer coupling. In particular, Jones discusses the role of quantized electronic states in magnetic multilayers and their dependence on the magnetic structure of the multilayer.

In the following paper, Himpsel et al. discuss the experimental observation of such quantum-well states in (100) Co/Cu using spin-polarized photoemission spectroscopy and the relationship of such states to both interlayer coupling and GMR.

While sputter-deposition techniques account for many of the major developments in the field of magnetic multilayers in recent years, MBE continues to be a valuable technique for the growth of certain single-crystalline systems.

Farrow discusses the role MBE has played in developing an understanding of certain aspects of GMR and oscillatory interlayer coupling. Much work was applied to understanding why single-crystalline (111)-oriented Co/Cu multilayers grown by MBE show no GMR, whereas polycrystalline (111)-textured Co/Cu multilayers grown by sputter deposition show very large GMR values. For many years this was a puzzle which was addressed by many groups. It is now generally believed that imperfections in the MBE-prepared multilayers account for this behavior.

The use of advanced materials growth techniques, such as the use of surfactants to promote smoother films, has finally led to the observation of oscillatory interlayer coupling in MBE-grown (111) Co/Cu. These studies emphasize that subtle structural modifications of magnetic multilayers can lead to dramatically altered properties.

Nesbet reviews theories of GMR and, in particular, discusses models of GMR based on realistic band structures of the metals involved.

These models demonstrate the importance of the scattering at the interfaces between the ferromagnetic and nonferromagnetic layers in GMR. The importance of interface scattering to GMR, originally demonstrated in experiments which modified the interfaces, is now widely accepted and is of particular technological significance.

In recent years there have been many advances in experimental techniques for probing aspects of the structure and magnetism of magnetic layers and multilayers. Stöhr and Nakajima discuss the use of advanced synchrotron-based techniques to probe, with elemental specificity, not only the magnetic moments of the ferromagnetic layers in magnetic multilayers but also the magnetism of the nonferromagnetic layers.

Copper layers in Co/Cu multilayers become magnetically active by virtue of their proximity to the ferromagnetic Co layers. By contrast, in Fe/Cu structures, the ferromagnetic Fe layers can become magnetically inactive because of the structural influence of the Cu layers.

In the paper which follows, Sun et al. discuss recent developments in an entirely different class of materials: the manganite perovskites. These materials show extremely large changes in resistance in magnetic fields, but typically at temperatures well below room temperature and in extremely large magnetic fields. The origin of their magnetoresistance, commonly referred to as colossal magnetoresistance, is quite different from the GMR of magnetic multilayers. In my opinion, it is unlikely that such materials will be useful technologically because of their great complexity as well as the strong temperature dependence of their properties. In contrast to these materials, GMR multilayered structures have already found important technological applications.

In the final paper of the group, Tsang et al. discuss the use of specially engineered GMR structures for hard disk drive read-head applications. As mentioned above, GMR multilayered structures can display very large magnetoresistance values, but at relatively large magnetic fields.

In a magnetic recording disk drive, information is stored as magnetized regions within thin magnetic layers (which themselves can be rather complicated multilayered thin-film structures) deposited onto aluminum or glass platters. The regions are magnetized longitudinally within the plane of the platter either parallel or antiparallel to the circumference of the platter. It is the transitions between these magnetized regions which are detected by a magnetic recording read head via the magnetic flux emanating from the transition regions. The strength of the magnetic field at the read head is only a few tens of oersteds. Thus, a field-sensing device capable of detecting such small magnetic fields is required.

GMR magnetic multilayered structures can be engineered to be sensitive to such small fields with greater sensitivity than conventional ferromagnetic metals. For example, this can be achieved by forming simple sandwiches of two ferromagnetic metals separated by a thin layer of Cu perhaps 20 Å thick.

One of the ferromagnetic layers is magnetically hardened so that for the range of magnetic fields to which it is subjected, the magnetic moment of that layer is fixed. This can be achieved by coupling that ferromagnetic layer to a thin antiferromagnetic layer using the phenomenon of exchange biasing. In addition, one takes advantage of the fact that whereas the GMR effect falls off very slowly with increasing separation of the ferromagnetic layers for Cu spacer layers, the magnitude of the interlayer exchange coupling decreases much more rapidly with increasing Cu thickness. Thus, the intrinsic magnetic coupling through Cu layers 20 Å thick is weak.

The moment of the second ferromagnetic layer in the GMR sandwich rotates in the presence of the magnetic field from the transition region and thus causes a change in the resistance of the sandwich. The variation in the resistance of the sandwich is used to detect the transition region between the magnetic bits, and thus the stored data can be recovered.

This type of GMR sandwich has been termed a spin-valve GMR sandwich. Tsang et al. discuss the optimization of GMR multilayered structures for high-density magnetic recording read-head applications. For such applications, very thin ferromagnetic sense layers are needed. One of the main reasons why GMR structures provide a greater signal for these applications than conventional ferromagnetic alloys is the predominant interfacial origin of GMR, which typically results in a larger GMR signal for thinner ferromagnetic layers.

The papers included in the GMR-related portion of this issue cover just a small subset of the work carried out within IBM over the past few years on magnetic multilayers and related structures. This work is itself just a small subset of that carried out within the scientific community. It is perhaps remarkable that within less than a decade, magnetic multilayers have evolved from a scientific curiosity to become materials of significant technological importance.

This class of magnetic multilayers has revealed a considerable number of scientific insights. In addition, the study of its properties has led to significant developments in the preparation of magnetic multilayered structures with layers just two or three atomic layers thick.

Such structures can now be prepared routinely and with sufficient reproducibility for commercial applications. Advances in structural, magnetic, and electronic characterization have helped to motivate continuing interest in such structures. In the future, studies seem likely to evolve to the study of lower-dimensional magnetic nanostructured materials patterned or structured into lines or dots.

MAGNETIC NANOMATERIALS IN DATA STORAGE APPLICATIONS

Magnetic nanoparticles are useful for a wide range of applications from data storage to medicinal imaging. The large-scale preparation of FeCo nanoparticles boosts this potential.

Nanoparticles offer exciting opportunities for technologies at the interfaces between chemistry, physics and biology. Their appeal stems not only from their use as single particles, but also from their potential to form self-organized films and solids.

Nanomagnetic systems are an important part of today's technology and have potential applications in future new devices. The applications of nanomaterials as patterned media for hard disks, miniature magnetic sensors using GMR effect and magnetic memory cells which are being developed for fast and dense solid state memory will be outlined. The areal density of magnetic recording has been increased over the past few years at 60 % per annum due to concurrent optimization of successive generations of thin film heads, miniaturization of the planar write head and improvement of Co-based in-plane thin film media.

Figure: ***Magnetic nanoparticles: Applications***

The hard disks having a greater storage density have been developed. However, the effects associated with the finite grain size of magnetic alloys used for storage, limit the miniaturization. As a result a new type of storage media has been proposed made from a closely-packed array of nanomagnets patterned onto a hard disk. A conventional hard disk is constituted by a thin film of granular Co-alloys. The thin films are constituted from exchange decoupled grains, each grain reversing individually their magnetization. Intergranular exchange coupling is suppressed by adding a non-magnetic element (particularly Cr) which segregates at the grain boundaries. In order to increase the storage capacity, it is necessary to reduce the size of the grains and to ensure a minimum number of grains per bit, to limit statistical noise. It was suggested that for obtaining low particle noise contribution, about 1000 grains are needed in one bit cell. For example a bit cell in the Co-based medium used for the IBM (0.184 Gbit cm-2) consists of about 1900 grains with a grain size of about 20 nm. Thus, for a 1.5 Gbitcm-2 thin film medium, a grain size of about 10 nm is necessary. In this case there will be about 600 grains in a 1 bit area.

The reduction of the size of grain is limited to E" 10 nm, since the grains having smaller dimensions are superparamagnetic. Smaller grains can only be used if their switching field is increased. This may be possible by increasing the coercivity of Co alloys, up to 0.28 MAm-1, by incorporating Ta, Pt. Coercivities up to 0.8 MAm-1 would be possible by using permanent magnet materials (Nd-Fe-B, Sm-Co). In this case the grain size can be reduced up to 2.5 nm but in this case it is not possible to use write heads, available today, in which the write fields are limited to maximum 0.4 Mam-1.

As mentioned already, higher storage densities, in the range (16 – 80) Gbitcm-1, could be reached by using patterned media. The collection of exchange coupled grains in this case is replaced by an array of nanomagnet bits (10 – 20 nm sizes) physically isolated from their neighbours and able to be switched "independently".

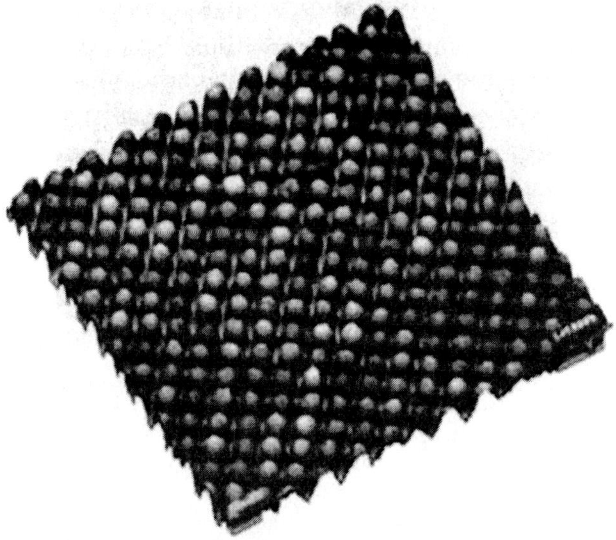

Figure: ***Sun is intruiged by the potential applications***

The bit could be oriented either in the plane of disk or perpendicular to it. There are plotted three types of patterned media with bars in the plane of the disk (a), perpendicular pillars (b) and cones (c). The perpendicular configuration leads to a more dense packing, but in-plane configuration allows a more conventional reading and writing technology to be used. Nanomagnet bars of Co film with dimensions 40x25x200 (nm), in plane, have a storage density of 4 Gbitcm-2. The average switching field is E" 0.1 MAm-1. We note that the packing density is limited by the magnetostatic interactions between elements.

Giant Magnetoresistance Sensors

Magnetic memory elements are being developed as a new type of non-volatile random access memory (RAM). As mentioned before, in certain applications magnetic RAM could replace silicon devices based on CMOS technology, particularly where the resistance to radiation is important. The operation of magnetic memory cells is based on the giant magnetoresistance effect (GMR) in spin valves or spin tunnel junction structures and special properties of nanomagnets as storage elements for binary data. Giant magnetoresistance was first discovered in Fe-Cr superlattices with the non-magnetic Cr layers less than 2nm thick, separating slightly thicker Fe layers.

The origin of the GMR effect is the difference in scattering of electrons of opposite spins (up or down) as they pass through a ferromagnetic materials. The GMR sensor could consist of two magnetic layers which have their magnetizations oriented parallel or antiparallel to each other. When the two layers are magnetized antiparallel, the spin-up and spin-down electrons will be scattered equally. When the layers are magnetized parallel, the spin-dependent scattering results in a lower resistivity for one type of electrons and a higher resistivity for the other type. The lowering resistivity for one type, result in a lower resistance for the structure, as a whole device. Thus, when the direction of magnetization direction is changed, multilayers structures, spin valves and spin tunnel junctions can be used as sensors.

This involves that they must be patterned into micron or sub-micron sized elements. Multilayers comprising of alternating series of magnetic - non-magnetic bilayers can produce large "R/R values, but requires a field of the order 10-40 kAm-1 to reverse the antiparallel coupling of layers and to saturate them in the low resistance state. Thus, the multilayers cannot be used as detectors of small fields. A more sensitive response is possible by using a spin valve structure. The spin valves have two soft magnetic layers (Ni-Fe) separated by a metallic spacer layer (Cu).

The GMR effect is generated by having a "pinned" layer held rigidly to a fixed magnetization direction and a layer in which the magnetization is free to rotate relative to the pinned layer when a field is applied. Thus, high resistance (antiparallel) and low resistance (parallel) states can be achieved. Spin tunnel junctions have two magnetic electrodes separated by an insulating tunnel barrier made from aluminium oxide. Unlike in spin valves, the current flows through the junction perpendicular to the layers. This increases the effect of spin dependent scattering on the electrons passing through the structure and leads to higher "R/R values than in spin valves. The interesting physical properties of magnetic nanostructures materials as well as their characterization will be largely presented in the following lectures. The analyze methods will be also described. In addition come of their technical applications will be outlined.

Magnetism is the underlying effect of many technical achievements provided by science and industry during the last decades. In today's research focusing on nanotechnology and electronic data storage, interest in magnetism even increased, and future attempts in micro-design require a deeper knowledge of magnetic properties on the nanoscale. In contrast to optical or acoustic phenomena, which in principle are accessible by human perception, magnetism cannot be directly sensed and its experimental investigation is restricted to indirect observation. This can be done by measuring the effect of magnetic order on other physical properties of a solid like electrical and thermal conductivity, elasticity, heat capacity, sound attenuation or lattice strains – all these properties show anomalies at the magnetic phase transitions.

Another important access to magnetism is given by magneto-optical (MO) effects, which relate the optical properties and the magnetic state of a solid. In the visible light region MO effects are widely used to measure the average magnetization of thin samples (thickness << wavelength of the probing light) and therefore has become a standard tool in thin- film magnetism research. Although MO effects in the visible-light region are quite small, refined optical methods not only yield sufficient contrast for distinguishing bits of opposite magnetization, but even allow one to observe details of magnetic domains in optical microscopes. Most importantly, MO techniques are ideally suited for studying magnetization reversal processes in external magnetic fields, which is impossible with, e.g., electron microscopes involving slow cascade electrons that are strongly affected by the Lorentz force.

Besides all merits, conventional MO techniques lack element specificity, simply because electric dipole transitions in the visible-light region mainly take place between delocalized conduction-electron band states. This has become a severe limitation in analyzing compounds containing several magnetic elements, as required for information storage or permanent-magnet nanostructures. In these advanced magnetic systems, lanthanide elements (mostly Tb) are used to achieve large perpendicular magnetic anisotropy, and large coercive fields.

Reliable element-specific information can be obtained by techniques that involve core-level transitions, e.g. by magnetic circular dichroism in x-ray absorption (XA). In the soft x-ray region, absorption signals are commonly detected by electron-yield methods, so that in general no large external magnetic fields can be applied, and hence only very few hysteresis loops have been recorded via electron yield. Alternative XA detection modes which do not employ electrons are either limited to specific samples that are suspended on transparent foils, or they suffer from saturation effects for all but highly diluted samples (fluorescence yield).

Thus, to probe the field-induced magnetization reversal of a compound in an element-specific way, it is useful to employ x-rays in *reflection* like in conventional magnetooptics.

Resonant enhancement of x-ray scattering cross sections at inner-shell absorption thresholds have been used for years to investigate the magnetic structure of lanthanide and actinide materials in the hard x-ray region (above 2 keV). The experimental demonstration of large changes in the specularly reflected x-ray intensity at the Fe L 2,3 threshold upon magnetization reversal initiated the ongoing search for magneto-optical (MO) effects in the *soft x-ray* region, and for ways to apply them to element-specific studies of heteromagnetic systems. Apart from their technological relevance in soft and hard magnetic films, lanthanide elements constitute a unique case in x-ray optics: the *4d* " *4f* electric dipole transitions (the so-called giant resonances) are among the strongest in the periodic table.

They give rise to high absorptivity and reflectivity, corresponding to a very short x-ray absorption length which in the course of the present work were found to be comparable or even shorter than the lanthanide *4d* " *4f* transition wavelength (*l H*" 8 nm).

This unique situation is quite different from previously studied transition-metal (Fe, Co,...) *2p* " *3d* transitions, where the attenuation lengths are much longer than the wavelength, giving rise to complicated interferences when samples are structured on the nanometer scale. Yet, in analyzing soft x-ray MO signals from thin films and multilayer systems with thicknesses comparable to the x-ray wavelength, previous investigations have shown that a comparison with model calculations (based on the Fresnel equations) is needed in order to extract layer-resolved sample magnetization profiles. Several experimental determinations of soft x-ray MO constants have been reported for ferromagnetic transition metals in the region of the L2,3 thresholds, but none so far for the lanthanide elements.

The purpose of this thesis is to investigate MO effects of lanthanide elements in the soft-x-ray region. The results are compared with visible-light MO data in order to discuss their mutual compatibility. Special attention is paid to the element-specific behavior of heteromagnetic systems. The elements Gd and Tb are chosen as examples because of their similarity in lattice constants but widely different magnetic behavior.

SCALABLE PARALLEL MICROMAGNETIC SOLVERS FOR MAGNETIC NANOSTRUCTURES

A parallel finite element micromagnetics package has been implemented, that is highly scalable, easily portable and combines different solvers for the micromagnetic equations. The implementation is based on the standard Galerkin discretization on tetrahedral meshes with linear basis functions. A static energy minimization, a dynamic time integration, and the nudged elastic band method have been implemented. The details of the implementation and some aspects of the optimization are discussed and timing and speedup results are given. Nucleation and magnetization reversal processes in permalloy nanodots are investigated with this micromagnetics package.

In various fields of computer aided engineering like structural analysis, fluid dynamics, and electromagnetic field computation, as well as micromagnetics the finite element method has been successfully applied. In particular, its flexibility in modeling arbitrary geometries has made it very popular. In the light of the importance of the microstructure of magnetic materials the finite element method has been chosen for the implementation of a micromagnetics package. There are several commercial and open source micromagnetics packages available, however all of them use the finite difference method. Our finite element micromagnetics package combines several unique features:

It is:

- entirely based on portable, free, open source software packages,
- highly portable to different hardware platforms, which range from simple PCs to massively parallel supercomputers,
- highly optimized and scalable,
- well integrated, combining static energy minimization, dynamic time integration, and nudged elastic band methods.

7

Nanotechnology Revolution

It may seem like science fiction, but the nanotechnology revolution is already underway. It's going to change our lives in ways we've only begun to imagine.

Nearly 40 years ago, the late Isaac Asimov wrote a sci-fi classic about a dramatic race against time and a technology so unbelievable that the author aptly titled his book Fantastic Voyage. Back in 1966, Asimov's scenario was sensational. In retrospect, however, it's his prescience that's amazing.

The book and the film it inspired tell the story of a desperate struggle between Cold War adversaries for control of a still-experimental technology called miniaturization, which reduces the size of very large objects by shrinking their atoms. A five-member medical crew undergoes extreme miniaturization and travels through the body of a professor (and communist defector) via a tiny submarine. The crew's mission is to destroy a life-threatening blood clot in the defector's brain, thereby saving his life and the intellectual capital that will ensure democracy's triumph.

Today, the technology Asimov envisioned in Fantastic Voyage is far closer to becoming fact than many people realize. A burgeoning new field called nanotechnology has sparked the scientific imagination and generated a surge of research activity across many disciplines. It's also instigated a high-stakes race for knowledge and dominance of the still-experimental technology. Many experts believe that nanotechnology will someday revolutionize science, medicine, and industry. Nanotechnology manipulates and controls the physical world's smallest, most basic components to obtain a desired result. The properties of a material-such as its mechanical strength, electrical conductivity, and complexity-are determined at the molecular level and are dependent upon the arrangement of its atoms. If they had the ability to manipulate and control the structure of materials at or near the scale of a nanometer-a billionth of a meter-researchers theoretically could invent and improve materials, technologies, and products to a degree that's almost unimaginable today. The most basic example of existing nanotechnology is the nanoscale molecular machinery inside the living cell. Each cell works like a machine, converting fuel to energy and pumping out proteins and enzymes as directed by its DNA. However, within the limitations of its existing structure, a cell can perform only certain highly circumscribed functions. Nanotechnology will break down these limitations and give humans unprecedented control over the material world. Nanotechnology's ultimate goal is the creation of ultrasmall computers and machines as well as advanced materials that can carry out vital engineering or medical tasks at the molecular level.

Advances in nanotechnology will affect a number of major industries, including electronics, medical science, pharmaceuticals, and computers. It could lead to better recording devices, faster computers, and better protective coatings.

Researchers also envision some amazing products and devices that smack of science fiction, such as an "invisible" airplane with a chameleon-like ability to blend into its surrounding environment, thanks to a special exterior coating. Someday construction materials may come equipped with built-in sensors that detect weather conditions and control the indoor climate of homes and offices automatically.

And a technology that was the stuff of science fiction just a few decades ago is a step closer to becoming reality at the University. Researchers here are working to create a gene gun, a nanoscale medical weapon that can enter the human body and fight diseased cells one at a time. The project is one of the University's most important nanotechnology accomplishments to date. "The 'U' has pockets of excellence in this field," says David Pui, Distinguished McKnight University Professor and director of the University's Particle Technology Laboratory, where the experimental gene gun is being developed. "Nanotechnology is inherently multidisciplinary, and a comprehensive university like ours provides excellent opportunities to conduct such research."

Still in its infancy, nanotechnology has captured the attention of researchers and world governments alike. However, when physicist Richard Feynman introduced the concept of nanotechnology in 1959, the idea of manipulating matter atom by atom received a cold reception. Serious research in the field didn't begin until the late 1980s and has only recently become a primary focus of researchers.

National governments have accelerated their support for nanotechnology research and development. In 1999, the U.S. government invested about $260 million in nanotechnology research, and Japan and Europe made similar investments. Last January, President Clinton declared a National Nanotechnology Initiative, and his FY 2001 budget request included a $227 million increase (84 percent) in government funding for research and development. The White House National Science and Technology Council recently created the Interagency Working Group on Nanoscience, Engineering, and Technology. Members from eight federal agencies are working closely with academic and industry leaders to explore nanotechnology's potential on a national level.

The National Science Foundation (NSF) currently is receiving proposals to fund nanotechnology research. Years of experimentation and research in nanotechnology have yielded promising results for the University, which is preparing a proposal for submission to the NSF.

University researchers are working on nanotechnology projects in micromachines, medical devices, computers, electronics, and a number of other fields. Others are exploring the world of microsystems and particle interaction.

All this research activity has moved the University to the forefront of the nanotechnology revolution, according to industry observers. The University also has won recognition for promoting interdisciplinary research collaborations among various academic units. The Institute of Technology, College of Biological Sciences, Academic Health Center, and College of Agricultural, Food and Environmental Sciences all are involved in nanotechnology experimentation and research.

For the hundreds of medical device, telecommunication, manufacturing, and biotechnology companies based in Minnesota, advances in nanotechnology hold the key to future innovation and economic success.

Last March the University hosted the Minnesota Nanotechnology Summit, which brought national experts together to discuss the impact and potential of nanotechnology. The summit also gave the University a chance to showcase its accomplishments.

According to University president Mark Yudof, the conference was a first step toward identifying how Minnesota's public and private sectors, together with the University, can capitalize on nanotechnology's potential and guarantee that the University remains among the leaders in the field.

"There are a lot of good things going on at the University of Minnesota," Pui says. "The nanotechnology summit was one of the first among research universities nationwide to promote interdisciplinary collaboration. The University's central administration and the deans are very supportive of this initiative."

On the cusp of a technological sea change, the world is likely to see a number of scientific breakthroughs in nanotechnology over the next 25 years. University researchers are working diligently to ensure that they're responsible for some of those fantastic journeys.

FUNDAMENTALS AND THEIR EXPLORATION

High hopes and unanswered questions accompany any emerging technology, and nanotechnology is no exception. For nanotechnology to fulfill its promise, researchers must find the answers to fundamental questions about how microsystems and particles interact with each other.

To fill that knowledge gap, some researchers are turning to mesophysics, part of the broader interdisciplinary field of nanoscience. The mesoscopic scale encompasses systems that are larger than single atoms but small enough that their properties differ radically from larger pieces of matter. An understanding of mesophysics is essential to achieving higher densities of microelectronic semiconductor-based circuits.

One such project is the focus of Leonid Glazman, McKnight Presidential Chair of Theoretical Condensed Matter Physics, who is considered one of the world's leading authorities in the field. Glazman's research explores the theory of electron transport and correlations in nanoscale devices.

In most conductors, electrons carry the electrical current through a wire, like a stream coursing through its channel. The width of a conventional wire is about ten million times the size of an electron, so the electron can pass through it with ease.

The quality of a wire as a conductor of electrons is defined by the wire's conductance. When wires become extremely small, however, unusual phenomena occur. Conductance breaks down when a wire's width is small enough to guide or control the passage of single electrons through the wire. Then the quantum properties of electrons- their ability to behave as waves, as well as the finite and universal value of electron charge-become important.

Thanks to advances in the semiconductor industry, it's now possible to produce small devices that morph continuously, ranging from a waveguide for quantum waves of electrons to islands that carry a tiny discrete charge produced by only a few electrons.

By studying the fundamental properties of these smaller mesoscopic systems, including the particle and wave properties of electrons, Glazman and other physicists are uncovering the fundamental physics behind nanotechnology.

Other University researchers are intent on learning how to control the size, composition, and other properties of nanoparticles. Andreas Stein, an associate professor of chemistry, is developing nanoporous solids that mimic the structures of natural porous minerals called zeolites. Functioning as molecular sieves, zeolites allow molecules of designated sizes and shapes to pass through and react inside their channels under very controlled conditions. Synthetic zeolites have various applications in industry. The petroleum industry uses them to increase the octane rating of gasoline, a mixture that contains molecules of various sizes. When gasoline is filtered through zeolites that have reactive channel walls, certain molecules in the gasoline react with the walls in a process that increases the octane levels. For the reaction to occur, however, the molecules must come in contact with the channel walls.

One aspect of Stein's research has dealt with modifying the surface reactivity of porous solids. "We want to go beyond merely replicating nature's three-dimensional zeolites," he says.

Typically, the channels or pores of natural zeolites are just large enough to accommodate small molecules, but researchers need porous solids that give them more flexibility and control. If researchers can vary the size and arrangement of the channels and pores, they'll have greater control over the types of molecules that can interact with the sieve-and over other properties of a given material.

It's already possible to design synthetic porous solids to accommodate larger molecules. "We can now prepare solids with multiple pore sizes and in a large variety of compositions," Stein says. "This [capability] has opened the door to new applications."

Other possible applications for modified porous solids include drug delivery, catalysis, and optical technology.

One of Stein's more recent projects involves the synthesis of photonic crystals, which don't absorb light of a given color. If a paint were created using photonic crystals that don't absorb the color green-the most intense color in the spectrum-then a car whose exterior was coated with that paint wouldn't absorb green light and would remain cooler when exposed to direct sunlight.

In the Reacting Flow and Nanoparticle Laboratory, Michael Zachariah is studying methods of controlling the morphology of nanoparticles grown from the vapor phase. "The idea, essentially, is that the properties of particles sometimes depend on their size," he says. "If you can control the size, you can control the property."

Zachariah, an associate professor of mechanical engineering and a member of the chemistry department's graduate faculty, says that the ability to manipulate particles would benefit nearly every manufacturing industry, especially in the areas of ceramics, electronics, and fiber optics.

Although they're exceedingly small, nanoparticles can contribute to problems on a global scale.

Peter McMurry, professor and head of the mechanical engineering department, has been studying atmospheric nucleation since 1972. Gas-phase chemical reactions cause atmospheric nucleation-the formation of particles of precipitate-which affects the earth's radiation balance and the formation of aerosol-induced haze.

Over the years, McMurry and his research team have made several important discoveries and developed instruments to detect the presence and measure the composition of nanometer-sized particles. One of these instruments has been commercialized and is now widely used. McMurry's group has conducted field measurements of atmospheric nucleation over the Arctic Ocean, Hawaii, Colorado, Atlanta, the Indian Ocean south of Tasmania, and at the South Pole.

Steven Girshick, professor of mechanical engineering, is the principal investigator of a study that tests nanostructured films for friction and wear-resistance. The three-year, $660,000 project is funded by the National Science Foundation. His research has led to new methods of depositing stronger and more efficient nanostructured coatings.

But depositing the nanoparticles can be tricky. As a flow carrying particles approaches a surface, very small particles follow the flow as it curves around the surface, like snowflakes that curve around a car's windshield. To deposit a coating of nanoparticles efficiently, the particles must be traveling at high velocity.

Girshick and his colleagues invented a patented process called hypersonic plasma particle deposition, in which nanoparticles nucleate in a plasma that's accelerated through a nozzle. The particles-about 20 nanometers in size-hit the targeted surface at a velocity of two kilometers per second, creating a dense nanocrystalline coating. According to Girshick, this type of coating would improve friction and wear-resistance in automotive shafts and ball bearings as well as in biomedical implants. "There is simply a huge number of possibilities," he says. "There's been a lot of work [being done] worldwide to make nanostructured coatings. This process is unique to Minnesota. It is a different approach."

PAVING THE WAY FOR SMALL MACHINES

Disposable satellite transmitters, inexpensive medical testing equipment and sensors for automatically tracking inventory or traffic patterns will become possible over the next 10 years through developments in nanotechnology, speakers at the Nanotech 2003 conference said Monday.

Nanotechnology-the science of making devices with features measuring less than 100 nanometers (or one-ten-millionth of a meter)-will let companies make smaller and cheaper products than ever before, which in turn will lead to new markets, according to Albert Pisano, a professor of engineering and computer science at the University of California at Berkeley.

Communications, for instance, could be greatly affected, Pisano noted. Right now, radios on satellites have to be hardened against radiation, an expensive process. At Stanford University, researchers have shown how small, unhardened radios can transmit and receive messages while in orbit. Although these smaller radios are subject to harm from radiation, they cost far less, so many could be mounted on a spacecraft to compensate for burnouts.

A market for tiny transmitters could also emerge within five years, out of a need to better monitor power lines, freeways and bridges for potential failures.

The movement will also mark a major turning point in mechanical engineering because most of these tiny systems will deal directly with the physical world, like pumps do, rather than getting information from electrical impulses, like microprocessors.

"Most people forget that a radio is two-thirds filter and that filters are mechanical," Pisano said. "Sensors and computing and communications can all be heavily miniaturized."

Batteries, too, could improve through breakthroughs in mechanical engineering. Currently, a kilogram of lithium ion, the same material used in notebook batteries, provides about 400 watt-hours of energy. A kilogram of methane provides 15,000 watt-hours of energy. The difference is that lithium ion is electrically charged, while methane provides thermal energy-that is, heat.

"Can you use thermal energy on such a small scale?" Pisano asked. Apparently so. Some researchers have already developed rotary engines that measure only microns across.

Of course, getting to miniature nirvana won't be easy. In the early '90s, futurists predicted a booming market for MEMS (microelectromechanical systems) similar to the devices described above. Difficulties in mass manufacturing, packaging and other problems kept the market from taking flight.

Pisano, though, noted that new technologies take time. About 15 years passed between the development of the transistor in the late '40s to the true development of the electronics industry. Several challenges need to be overcome, but the progress with nanotechnology makes MEMS more promising.

The first must-have applications for the types of MEMS products Pisano described will appear in about five years, he predicted, because the demand exists. Gillette, for instance, has spent heavily in the last few months on diminutive radio-frequency transmitters. A number of universities and private companies will be presenting papers on MEMS over the five-day conference here. Some of the major themes include microfluidics, or small machines that can test biological samples, and films or surfaces with sensory capabilities.

Other companies are presenting papers on progress in using chains of molecules to make processors or memory devices. Molecular computing will likely take much longer to develop, but early results are showing promise.

Molecular chips differ substantially from today's silicon processors, noted Phaedon Avouris, manager of nanoscience and nanotechnology at IBM's Watson Research Center. Carbon nanotubes exposed to the air, for instance, carry a positive charge. When hermetically sealed off, they carry a negative charge.

SMALL TRUTH OF NANOTECH

The drop-off was stomach-churning. In 2000, venture capitalists poured $100 billion into startups. Last year, they couldn't even reach $40 billion. So forgive them for latching onto nanotech as the uptrend. Headlines like "NANOTECHNOLOGY WINS OVER MAINSTREAM VENTURE CAPITALISTS" and "THE NEXT BIG THING IS VERY SMALL" are getting hard to avoid. Nano conferences are weekly events, crowded with VCs amped up about self-assembling machines and nanobots in your bloodstream. But what's really going on? While venture capitalists are happy to hype nanotech, they aren't exactly rushing in where angel investors fear to tread. Of the $5.6 billion put up by private capital sources in the first three months of this year, only $42 million went to very small tech, according to Steve Glapa, president of In Realis, a market consulting firm in Milpitas, California. Even putting all of these investments under the "nanotech" umbrella is a stretch. Advanced chemistry and materials science are more like it.

The reality of nanotechnology has less to do with self-assembling Ferraris than with powders and filters and goop. It's about taking a material like zinc oxide and grinding it down to grains the size of a few dozen nanometers (great for sunscreen). Or punching nano-size holes in screens so hydrogen atoms can slip through but water molecules can't (handy for fuel cells). Or laying down slurries a few atoms thick (good for hard drives). It's not a new industry so much as the outcome of decades of research. "People have been doing this sort of thing for a long time," says Glapa. "We just didn't always attach the nano prefix to it."

So why all the fuss from VCs now? Why the conferences, the speeches, the sound bites? Because it's great business. Attaching nano to a company name can instantly multiply the value. Or at least that's the idea. "VCs invest in companies where they see an exit strategy," says Glenn Fishbine, author of The Investor's Guide to Nanotechnology and Micromachines. "The more they can do to make nanotechnology attractive now, the better off they'll be when it's time to sell."

Unlike past technology booms, it's unclear whether VCs will ever have much of a role to play in the nanotech realm. Industrial labs at IBM, Hewlett-Packard, and 3M have invested hundreds of millions of dollars in scaling materials down to the atomic level, a gap that VCs are unlikely to close. "They are playing in exactly the same market," says Tom Theis, IBM's director of materials research. "They have no advantage over what big companies have looked at and rejected."

The structure of VC investing- seed money, ramp-up, IPO, exit- won't work easily with nanotech. It's an area where $800,000 government grants are readily available, often enough for a research scientist to get to a working prototype. The next step, bringing a product to market, will often take longer than VCs are willing to wait. So researchers with promising technologies are likely to bypass VCs altogether and sell directly to an established industrial player. That's the route Nano-Tex took when the Emeryville, California-based company developed a nanoscale technology to protect cotton from stains, then sold it to Burlington. "If someone has a really good memory technology, they won't go to a VC, they'll just go to IBM," says Fishbine. "Then they don't have to worry about marketing and manufacturing and distribution; they'll just take the royalty payments and buy land on Kauai." Not so fast, says Steve Jurvetson of Valley VC firm Draper Fisher Jurvetson. Before startups can partner with big industrial players or license intellectual property to them, they need backing from venture capitalists. At the same time, he adds, VCs need to relearn patience. During the Internet craze, the moneybags on Sand Hill Road got used to going from concept to IPO in as little as 18 months. Post-bust, VCs are returning to the standard three- to seven-year wait for a return. "I think we're on the long end of that for nanotech," says Jurvetson, adding that a dozen years isn't out of the question. "If the opportunity is big enough, if it's really going to change the world in a significant way, then it's worth the wait."

That thinking- and time horizon- may work for today's nano of chemistry and goop. But the VCs are luring investors with a grander vision: the idea that someday we might manipulate matter in fantastic ways, that we could build complex structures from the atomic level- much the way living things self-assemble from info stored in their DNA. Of course, that's all speculative stuff, and even nano evangelists like Richard Smalley say we're decades from realizing that dream, well beyond even Jurvetson's expanded VC timeline.

NEARING THE NANOTECHNOLOGY REVOLUTION

There's a lot of buzz-nanotechnology is "coming soon." But what is nanotechnology? Why doesn't anyone ever explain that? Well, it's not that easy. While experts agree about the size of nanotechnology-that it's smaller than a nanometer (that's one billionth of a meter) they disagree about what should be called nanotechnology and what should not. Looking back at the historical roots of nanotechnology helps us get a better grasp on what nanotechnology is and why it's important now, and how it will change the world in the future.

The story of nanotechnology begins in the 1950s and 1960s, when most engineers were thinking big, not small. This was the era of big cars, big atomic bombs, big jets, and big plans for sending people into outer space. Huge skyscrapers, like the World Trade Center, (completed in 1970) were built in the major cities of the world. The world's largest oil tankers, cruise ships, bridges, interstate highways, and electric power plants are all products of this era. Other researchers, however, focused on making things small. In the 1950s and 1960s the electronics industry began its ongoing love affair with making things smaller. The invention of the transistor in 1947 and the first integrated circuit (IC) in 1959 launched an era of electronics miniaturization. Somewhat ironically, it was these small devices that made large devices, like spaceships, possible. For the next few decades, as computing application and demand grew, transistors and ICs shrank, so that by the 1980s engineers already predicted a limit to this miniaturization and began looking for an entirely new approach.

As electronics engineers focused on making things smaller, engineers and scientists from an array of other fields turned their focus to small things-atoms and molecules. After successfully splitting the atom in the years before World War II, physicists struggled to understand more about the particles from which atoms are made, and the forces that bind them together. At the same time, chemists worked to combine atoms into new kinds of molecules, and had great success converting the complex molecules of petroleum into all sorts of useful plastics. Meanwhile geneticists discovered that genetic information is stored in our cells on long, complex molecules called DNA (about 2 meters of DNA is packed into each cell!) This and other work led to a greater understanding of molecules, which, by the 1980s, suggested entirely new lines of engineering research. So, the roots of nanotechnology lie in the merging of three lines of thinking-atomic physics, chemistry, and electronics. Only in the 1980s did this new field of study get a name-nanotechnology. This new name was popularized by physicist K. Eric Drexler, who pointed out that nanotechnology had been predicted much earlier, in an almost-forgotten 1959 lecture by Nobel Laureate Richard Feynman, who proposed the idea of building machines and mechanical devices out of individual atoms.

The resulting machines would actually be artificial molecules, built atom by atom. While the resulting molecule might itself be larger than a nanometer, it was the idea of manipulating things at the atomic level that was the essence of nanotechnology. But not only was this kind of manipulation impossible at the time, but few people had any idea why it would be useful to do it! With all the new research, however, Drexler revived Feynman's vision and helped introduce the general public to the basic concepts of nanotechnology.

Although nanotechnology dates from the 1950s, the biggest changes have occurred just in the past few years. In the late 1990s, research money began pouring in from corporate and government sources. In the space of just a few years governments around the world launched three major (and many other smaller) new research programs, including the National Nanotechnology Initiative in the U.S. and the nanotechnology branch of the European Research Area. Japan has its own huge nanotechnology program, with money coming from private industry and government agencies such as the Ministry of Trade and Industry.

THE BUILDING BLOCKS

Nanotechnology is a field that's just being established, and although there are big plans for the smallest of technologies, right now, most of what nanotechnologists have accomplished falls into three categories: new materials-usually chemicals-made by assembling atoms in new ways; new tools to make those materials; and the beginnings of tiny molecular machines.

Some of the primary building blocks in nanotechnology are buckminsterfullerenes (almost always known as buckyballs or fullerenes), which are clumps of molecules that look like soccer balls. In 1984 Richard Smalley, Robert Curl, and Harold Kroto were investigating an amazing molecule consisting of 60 linked atoms of carbon. Smalley worked these atoms into shapes he called "fullerenes," a name based on architect Buckminster Fuller's "geodesic" domes of the 1930s and first suggested by Japan's Eiji Osawa. Sumio Iijima, Smalley, and others found similar structures in the form of tubes, and found that fullerenes had unique chemical and electrical properties. Fullerenes became nanotech's first major new material. But what to do with them? Engineers turned their attention to finding some practical use for these interesting molecules.

While engineers thought about practical uses for fullerenes another discovery in search of an application was being made. In 1981 Gerd Karl Binnig and Heinrich Rohrer invented the scanning tunneling microscope or STM, which has a tiny tip so sensitive that it can in effect "feel" the surface of a single atom. It then sends information about the surface to a computer that reconstructs an image of the atomic surface on a display screen.

If that weren't amazing enough, a little later, researchers discovered that the tip of the STM could actually move atoms around, and Donald Eigler and a team at IBM staged a dramatic demonstration of this new ability, spelling out "IBM". Researchers believed they had a tool, the atomic force microscope (AFM), that could build things atom-by-atom. But, like the discovery of fullerenes, it remained to be seen if anything useful could actually be built this way.

The development of tools such as AFMs coincided with the introduction of very powerful new computers and software that scientists could use to simulate and visualize chemical reactions or "build" virtual atoms and molecules. This was especially useful for scientists working with complex chemical molecules, particularly DNA. Researchers recognized that the actions of DNA resembled some of the things nanotechnologists were now calling for-the use of molecules to construct other molecules, the self-replication of molecules, and the use of molecule-size mechanical devices. Perhaps DNA (or its cousin, RNA) could be modified to create the first nanomachines?

Geneticists had already found ways to use DNA taken from bacteria to make a nano-scale replicator used for scientific research. By modifying some of the chemical reactions that take place in natural DNA, genetic engineers had figured out a way to make copies of nearly any DNA molecule they wanted to study. But with the computers and tools available to them by the 1990s, they began using DNA or DNA-like molecules to do other things-like construct new chemicals or tiny machines. Many researchers began investigating ways to make proteins-the components from which DNA is made-that would perform useful tasks, such as interacting with other materials or living cells to create new materials or perhaps attack diseases. One of the first breakthroughs was Professor Nadrian Seeman's demonstration of a tiny "robot arm" made from modified DNA. While the arm could not yet really do anything useful, it did demonstrate the concept.

Meanwhile, electronics researchers approached nanotechnology from another direction. Since 1959, engineers had etched and coated silicon chips using a variety of processes to make integrated circuits (ICs). The transistors and other chip elements reached nano-scale in the late 1990s. They also used these same techniques to develop the first micromachines-microscopic devices with actual moving parts. Some of the early versions of these were simply intended to demonstrate the process without doing anything particularly useful, such as a tiny guitar with a string that could be plucked using an atomic force microscope. But in the late 1980s these began to be commercialized as machines-on-a-chip, or micro-electrco-mechanical systems (MEMs), which combine ICs and tiny mechanical elements. However useful MEMs are, most engineers feel that the techniques used to make ordinary ICs will never be refined enough to make true nanotechnologies. For that reason, engineers are now concentrating on discovering entirely new ways to make ICs, building them from the ground up rather than cutting and etching "bulk" silicon slices.

With the appearance of protein-based chemistry and other techniques in the 1990s, researchers began looking both for practical uses for nanotechnology and new ways to make nano-molecules or micromachines. A different but related problem was that of making nanomolecules in large numbers. A single nanomachine or nanocircuit for example, would not be able to do enough work to make a difference in the real world-thousands or millions might be needed.

Engineers needed ways to turn out their nanomachines in huge numbers, and so they began looking for a way to make a nano-scale machine or molecule that would assemble other nano-scale machines or molecules. K. Eric Drexler called it a "self assembler," and scientists believe that it will be one of the keys to making certain kinds of nanotechnology useful and practical. To date, very few practical nanotechnologies and no self-assemblers have been used outside the laboratory.

THE PRESENT STATUS

Nanotechnology is a science in its infancy, but that doesn't mean it hasn't been put to use. What exactly has been accomplished in nanotechnology so far? In general, all of today's practical nanotechnologies are those using nano-size particles of various materials, or nanometer-size features on integrated circuits (ICs), rather than the complex molecular machines that engineers first envisioned. These current nanotechnologies are still made by "top down" methods (like those used in conventional chemistry and IC manufacturing), rather than the largely unproven "bottom up" techniques predicted by nanotechnology's boosters.

Many current nanotechnologies, for example, consist of the ever-shrinking transistors, interconnecting wires, and other features on digital ICs. As of 2005, some integrated circuits now have transistors that measure about 50 nanometers across-well inside the accepted size-based definition of nanotechnology. But chips are still made using advanced versions of the lithographic processes developed in the 1950s, which layer on materials and then carve away at them to form the electronic circuits. They are not, in other words, constructed molecule-by-molecule from the bottom up.

However, chip manufacturers point out that when working with extremely small circuit elements, the behavior of electrons changes, so entirely new principles are at work. Also, there is at least one new chip with a somewhat different claim to being "nanotechnological." This is IBM's "Millipede" memory chip, which draws its inspiration directly from the Atomic Force Microscope (AFM). Electronics manufacturers can also point to the latest generation of high-density computer hard drives, which have extremely thin coatings of just a few atoms' thickness applied to the surface of the disc by a process called chemical vapor deposition.

While such nanochips are beginning to appear in greater numbers, most of us more often encounter applications of nanotechnological materials that are made in "bulk" form and added to other products. By far the best known of these are the controversial "nanotechnology" trousers introduced by The Gap and Eddie Bauer stores in 2005. These were simply ordinary cotton pants, treated with nanoparticles of a new, stain-resistant chemical that attached itself to the cotton molecules.

Carbon nanotubes, which can now be made in large quantities at relatively low cost by companies like Hyperion Catalysis International Inc., are being incorporated into a wide range of other products. Because the fibers conduct electricity very well, Hyperion was able to mix them into plastic compounds, which auto makers can then mold into parts that conduct electricity. This is useful for preventing static electricity charges from building up on parts such as plastic fuel system components, where the static can eventually damage them or, in some cases, cause a spark.

Nanotubes mixed into plastics are very strong and light, and have been used to make car body components, tennis rackets, and other items. They have also been used to improve battery performance, and may some day be used in other technologies that traditionally used ordinary carbon or metals to conduct charges. Infineon Technologies in Germany, for example, has demonstrated the use of the tubes to connect components on microchips. In 2002 they showed how nanotubes could be used to replace ordinary metal wires allowing them to carry more current but taking up less space. That would result in computer chips that can pack more circuits into less space; one of the longstanding goals of chip designers. One very useful new material is the semiconductor quantum dot. While not used in electronic circuits, quantum dots are nonetheless made from the same silicon used in computer chips. These tiny bits of material are coming into widespread use in experimental biology and, in a limited way, in medical diagnosis. The dots can be coated with certain chemicals, which are specially formulated so that they bind themselves to particular things-such as RNA, cell walls, or other types of molecules found in cells. One interesting application of this technology is its use in analyzing DNA material taken from the body. These DNA "scanners," first introduced commercially by Matsushita Corporation, combine integrated circuit technology and quantum dots to analyze genetic material much more rapidly than was possible before, and may lead to more rapid assessment of diseases. A second use of coated quantum dots is injecting them into the body, where they circulate until they come in contact with whatever type of cell their coating is designed to attach itself to. Then when a powerful infrared light source is shone on the body, it penetrates the flesh, illuminates the massed quantum dots, and the reflections can be detected to provide a "live" picture of an organ, muscle, cancerous growth, or other internal part without the need for surgery. Unfortunately, not all of these quantum dots are suitable for injection into a living human body, and some are even poisonous, but bioengineers are working around that problem.

Even with these real-world applications, the current uses of nanotechnology (other than nano-size particles of various materials) remain very limited. In fact, several once-promising nanotechnology based systems introduced commercially in the 1990s did not meet with success, such as the nanotube-based Field Emission Displays proposed as competitors to other flat-panel information displays. However, researchers are rapidly making progress toward what some think of as true nanotechnologies-self-assembling, molecule size machines to perform all sorts of tasks (including manufacturing the nano-size materials made by other methods today). The nanotechnological future, we are told, is right around the corner.

NANOTECHNOLOGY IN COMING FUTURE

The future of nanotechnology is largely a question mark. Futurists say we are entering a new era, somewhat like the Industrial Revolution of the 18th and 19th centuries. That revolution changed nearly everything about the way people lived. But no one at that time could have predicted how those changes would unfold. Could we be on the brink of another very rapid period of profound technological and social change?

The nanotechnological revolution, if it occurs, will be just as unpredictable in the long-term, but scientists and engineers have laid out some pretty fantastic forecasts for the near-future. For example, some see great promise for the use of nanotubes in super-strong materials. Even though the plastic composites made today using relatively short nanotubes are not yet much stronger than earlier types of composites, long nanotubes are expected to be used for extraordinary applications like the proposed "space elevator." This system would replace rockets for the transport of payloads and people into earth orbit.

Another major area where nanotechnologists predict stunning changes is in medicine. Imagine a world where no one gets seriously ill, grows older, or even dies (until they want to). That is what the prophets of nanotechnology say is in store for the 21st century. Today's nanotechnologies used in medicine offer only modest benefits, such as the ability to target diseased or cancerous cells, making them easier to locate.

In the near future, engineers tell us, that will change. Tiny molecular machines, perhaps based on complex, branched molecules called "dendrimers" will be injected into the body not only to locate cancers but also to find and repair cells damaged by disease or aging. Livers and hearts damaged by natural wear-and-tear, inherited diseases, poor nutrition, or alcoholism will be fixed or even replaced. Genetically based ailments such as Alzheimer's will be cured by replacing the faulty genes.

Some futurists have predicted that the most profound changes will be the result of the introduction of molecular assembly "factories," perahps even small small enough to fit on a desktop. These would, some say, make it possible for virtually anyone to design and build virtually anything, using nanorobots or perhaps a new technology called "nanoink," created by NanoInk founder Chad Mirkin. Nanotechnology researchers like K. Eric Drexler and Ralph Merkle say that this could be done by imitating and improving upon the "manufacturing" that DNA accomplishes inside the body. In 1999, researchers such as Nadrian Seeman at New York University demonstrated the principle of using modified DNA molecules to build a tinymachine, and somewhat later Nanoink founder Chad Mirkin had demonstrated building up nanostructures by depositing layers of materials on a substrate.

These and other experiments have also led researchers to believe that they will eventually be able to assemble circuits atom-by-atom in order to create the next generation of computer chips. The circuits on these chips will be much smaller than what is currently possible, and will enable the building of much more powerful computers. What difference will that make? Some, like engineer Raymond Kurzweil, think that computers will have personalities and be as smart as humans within 20 years. We may even be able to "download" our own personalities into computers, to become virtual humans. With nearly unlimited computing power, programmers are sure they could create software that completely blows away anything possible today.

Not surprising, these amazing predictions have inspired fear as well as wonder. Environmentalists and others point out that nanotechnology may bring with it unexpected dangers. The nanomaterials being made today, like fullerenes, are often in the form of extremely small particles. Even when these particles are made from common materials like carbon, they may interact with the human body or the environment in ways that are unlike those of natural particles of the same materials. Some say that allowing nanoparticles to be included in products ingested or applied to the body may pose health risks for consumers. Others predict that nanotechnology may get out of control, causing a huge man-made disaster. Eric Drexler and others, such as computer engineer Bill Joy of Sun Microsystems, warned in 2000 that self-replicating machines might run amok if they escape into the environment, competing with natural bacteria, plants, and people for natural resources. Then, in 2002 the public's awareness of nanotechnology-the bad side of nanotechnology-was greatly expanded when author Michael Crichton published his best-selling novel Prey, about tiny, self-duplicating nanorobots that band together to try to take over the world.

Whether public fears are founded in fact, it is true that the future of nanotechnology has inspired as much caution as optimism. Recently, in response to public outcry, researchers such as Dr. Vicky Colvin of Rice University have begun evaluating the risks and rewards of current nanotechnologies. Colvin and other engineers believe that, with wisdom, they can bring the wonders of nanotechnology into being while avoiding the pitfalls.

8

Exploring Nanotechnology in Cancer

In the fight against cancer, nanotechnology introduces unique approaches to diagnosis and treatment that could not even be imagined with conventional technology.

Figure: ***Nanotech Cancer-Fighting Robots.***

Nanotechnology offers the unprecedented and paradigm-changing opportunity to study and interact with normal and cancer cells in real time, at the molecular and cellular scales, and during the earliest stages of the cancer process. Through the concerted development of nanoscale devices or devices with nanoscale materials and components, the NCI Alliance for Nanotechnology in Cancer will facilitate in near future their integration within the existing cancer research infrastructure. The Alliance will bring enabling technologies for:

- Imaging agents and diagnostics that will allow clinicians to detect cancer in its earliest stages
- Systems that will provide real-time assessments of therapeutic and surgical efficacy for accelerating clinical translation
- Multifunctional, targeted devices capable of bypassing biological barriers to deliver multiple therapeutic agents directly to cancer cells and those tissues in the microenvironment that play a critical role in the growth and metastasis of cancer
- Agents that can monitor predictive molecular changes and prevent precancerous cells from becoming malignant
- Novel methods to manage the symptoms of cancer that adversely impact quality of life
- Research tools that will enable rapid identification of new targets for clinical development and predict drug resistance

Use of Nanotechnology in Cancer

Nanoscale devices are somewhere from one hundred to ten thousand times smaller than human cells. They are similar in size to large biological molecules ("biomolecules") such as enzymes and receptors. As an example, hemoglobin, the molecule that carries oxygen in red blood cells, is approximately 5 nanometers in diameter. Nanoscale devices smaller than 50 nanometers can easily enter most cells, while those smaller than 20 nanometers can move out of blood vessels as they circulate through the body.

Because of their small size, nanoscale devices can readily interact with biomolecules on both the surface of cells and inside of cells. By gaining access to so many areas of the body, they have the potential to detect disease and deliver treatment in ways unimagined before now. And since biological processes, including events that lead to cancer, occur at the nanoscale at and inside cells, nanotechnology offers a wealth of tools that are providing cancer researchers with new and innovative ways to diagnose and treat cancer.

Nanotechnologies

Work is currently being done to find ways to safely move these new research tools into clinical practice. Today, cancer-related nanotechnology is proceeding on two main fronts: laboratory-based diagnostics and in vivo diagnostics and therapeutics.

Nanotechnology and Diagnostics

Nanodevices can provide rapid and sensitive detection of cancer-related molecules by enabling scientists to detect molecular changes even when they occur only in a small percentage of cells.

Nanotechnology and Cancer Therapy

Nanoscale devices have the potential to radically change cancer therapy for the better and to dramatically increase the number of highly effective therapeutic agents. Nanoscale constructs can serve as customizable, targeted drug delivery vehicles capable of ferrying large doses of chemotherapeutic agents or therapeutic genes into malignant cells while sparing healthy cells, greatly reducing or eliminating the often unpalatable side effects that accompany many current cancer therapies.

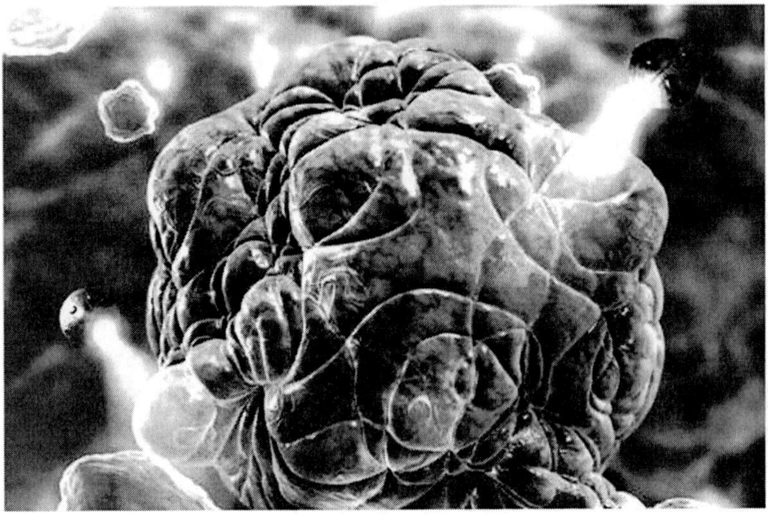

Figure: ***Nanotechnology takes on cancer. New technology has been forward that claims***

Nano Cancer Fix

A new cancer treatment could kill cancer cells without the painful side effects of chemotherapy. Using nanotechnology, scientists may now be able to destroy tumors without killing healthy cells at the same time.

Nanoshells

For people with cancer, chemotherapy, with its painful side effects, is often the worst part of the treatment process. What if there was a noninvasive cancer treatment that could target a tumor and leave the healthy tissue alone?

Now, nanotechnology may be on the way to providing just that. Naomi Halas, professor of chemistry and electrical and computer engineering at Rice University created tiny particles called nanoshells, which are 1/1000th the size of a human hair. "If you could see them they would look a lot like malted milk balls, except they are made of different materials," says Halas, "The crunchy core would instead be made of glass and the chocolate coating would instead be made of gold. And that's very important geometry because being able to control the relative size of the two parts of the nanoshell, the core and the shells, allows us to tune the optical properties of that nanoparticle such that we can control what wavelength of light it absorbs, what color of light it absorbs."

Halas reported in the journal Cancer Letters that the nanoshells' ability to capture light and convert it to heat has, in lab tests on mice, destroyed tumors. She and Rice biomedical engineer Jennifer West coat the nanoshells with a substance that binds them to cancer cells, and inject them into tumors. A harmless near-infrared light heats only the nanoshells, which convert the light into heat that destroys tumor cells, without affecting healthy tissue near the tumor.

Halas and West studied 25 mice with tumors that were divided into three groups: one that received no treatment; one that received saline injections and three minutes of exposure to the near-infrared light; and one that received nanoshell injections and a laser applied on the skin above each tumor.

While tumors in the other two groups of mice continued to grow, the tumors in the group that got the nanoshell treatment disappeared within ten days, and those mice remained cancer-free afterwards. "So far we have never seen any adverse reaction to nanoshells in animals," says West. "We've done a lot of testing at this point and all of the results to date look like they're very safe, non-toxic, and bio-compatible."

Halas says nanoshells would have virtually no side effects because they're not interacting with the organism. "They only are active once they're in place at the tumor site and once one shines light through the skin to the actual tumor site and heats that localized region," she says. "So we feel that is something might make a huge difference in terms of the invasiveness-not just the side effects but the overall lack of invasiveness of this sort of treatment. When nanoshells are directly injected into the bloodstream, the body doesn't recognize them. They're coated with specific molecules that essentially make them stealth particles, so they're not recognizable as a foreign object. Also...they're gold and gold is not recognized as a foreign substance by the human body. If we compare nanoshell-assisted cancer therapy to conventional cancer therapy for example, like chemotherapy, there's a very high likelihood that the type of side effects that people see in conventional therapies would be greatly reduced."

NCI Announces Major Commitment to Nanotechnology for Cancer Research

The National Cancer Institute (NCI) announced today at a media briefing a new $144.3 million, five-year initiative to develop and apply nanotechnology to cancer. Nanotechnology, the development and engineering of devices so small that they are measured on a molecular scale, has already demonstrated promising results in cancer research and treatment. "Nanotechnology has the potential to radically increase our options for prevention, diagnosis, and treatment of cancer," said Andrew von Eschenbach, M.D., director of the National Cancer Institute. "NCI's commitment to this cancer initiative comes at a critical time. Nanotechnology supports and expands the scientific advances in genomics and proteomics and builds on our understanding of the molecular underpinnings of cancer. These are the pillars which will support progress in cancer." To carry out this initiative, the NCI, part of the National Institutes of Health, is forming the NCI Alliance for Nanotechnology in Cancer, a comprehensive, integrated initiative encompassing researchers, clinicians, and public and private organizations that have joined forces to develop and translate cancer-related nanotechnology research into clinical practice. "The Alliance lays out a process to safely accelerate the application of nanotechnology to cancer research," said NCI Deputy Director Anna Barker, Ph.D. "Central to this initiative will be multidisciplinary partnerships involving physicists, biologists, clinicians, engineers, and other experts that can translate knowledge on cancer and nanotechnology into clinically useful products." The new NCI Alliance for Nanotechnology in Cancer is one of the first steps in implementing the Cancer Nanotechnology Plan, which was developed over the past 18 months with the input of a broad cross-section of the cancer research and clinical oncology communities. The NCI Alliance consists of four major program activities:

- o Centers of Cancer Nanotechnology Excellence (CCNEs): The primary goal of the CCNEs is to integrate nanotechnology development into basic and applied cancer research. Each center will be affiliated with a NCI Comprehensive Cancer Center, university, or research center of engineering and physical science. By leveraging existing NCI resources, these centers will bridge gaps in the development pipeline from materials discovery to preclinical testing.
- o Multidisciplinary research teams: Given the multidisciplinary nature of nanotechnology research, investigators with basic science and clinical backgrounds will require training to optimize the development and translation of nanotechnologies toward clinical oncology applications. The NCI will initially use existing career development mechanisms to direct talent to this area, create incentives for cross-disciplinary research, and foster collaboration through training.
- o Nanotechnology platforms for cancer research: Over the next five years, investigator-initiated and directed project research will be supported in six key programmatic areas: molecular imaging and early detection, in vivo imaging, reporters of efficacy (e.g., real-time assessment of

treatment), multifunctional therapeutics, prevention and control, and research enablers (opening new pathways for research).

- o Nanotechnology Characterization Laboratory (NCL): The NCL will perform and standardize the pre-clinical characterization of nanomaterials developed by researchers from academia, government, and industry. The NCL will serve as a national resource and knowledge base for cancer researchers, and facilitate the accelerated regulatory review and translation of nanomaterials and devices into the clinical realm.

The NCI recently signed a memorandum of understanding and an interagency agreement with the National Institute of Standards and Technology to partner with the NCI in this characterization and standardization effort. The NCI will also be working to expand collaborations with the U.S. Food and Drug Administration (FDA) to help define the critical pathway for nanotechnologies to reach the clinic.

Among the key components of the Cancer Nanotechnology Plan are milestones to measure success over two time periods. Within the first three years, the plan calls for acceleration of projects that hold promise for near-term clinical application. After three years, the Alliance will focus on developing solutions to address more difficult technological and biological problems that have the potential to impact detection and treatment.

"We are already seeing how nanotechnology is transforming our ability to translate research advances into clinical advances," said Samuel Wickline, M.D., Professor of Medicine, Physics and Biomedical Engineering at Washington University, St. Louis, Mo., and NCI grantee for nanotechnology research. "The possibilities are enormous for finding very small cancers far earlier than ever before and treating them with powerful drugs at the tumor site alone, while at the same time reducing any harmful side effects. This initiative will allow us to explore using this technology to its full potential." Recent advances in cancer treatment involving nanotechnology include:

- Liposomes, the "first generation" of nanoscale drug delivery devices, were developed to deliver anticancer therapeutics directly at tumors. Specifically, liposomal doxorubicin is being used to treat certain forms of cancer, while liposomal amphotericin B treats fungal infections often associated with aggressive anticancer treatments.
- Recently, a nanoparticulate formulation of the well-known anticancer compound taxol was submitted to the FDA as a new treatment for advanced-stage breast cancer.

Other clinical applications of nanotechnology have focused on identifying cancer in its earliest stages, visualizing development of the disease, delivering improved therapy to increase the effectiveness and reduce side effects of drugs, and capturing early signals of drug efficacy.

THE CANCER NANOTECH

Think small-really small. Nanotechnology employs devices with dimensions of one to 1,000 nanometers. To put this in perspective, consider the following: a nanometer is 1/80,000th the width of a human hair; it is the length of 10 hydrogen atoms placed end to end; and it is less than one third the height of a single twist on a strand of DNA. At these sizes it is no wonder that scientists have seized upon nanotechnology for myriad medical applications. After all, the human body is built upon a foundation of nature-made nanostructures-genes, proteins, cells-that may be best approached on their own scale. Cancer is the key area of medical nanotechnology research. In the not too distant future, dozens of intriguing nanodevices such as the nanotubes at left may transform cancer diagnosis, treatment, and prevention.-Lexi Krock

Nanotubes

Nanotubes are hollow cylinders made of carbon atoms. Doctors could someday use them as miniscule syringes for injecting cells with drugs from within the body, or as nanoscale diagnostic probes the patient would never feel. They can also be filled and sealed, forming test tubes or potential drug delivery devices. Here, an array of nanotubes engage in what their creators at the Oak Ridge National Laboratory call "impalefection," the capturing of genetic materials by impaling cells (a hamster's ovary cells in this case). In the future, nanotubes could help identify DNA mutations associated with a risk of cancer.

Nanowire

This glowing silica nanowire is wrapped around a single strand of human hair. It looks delicate-it is about five times smaller than a virus-but it is several times stronger than spider silk. Researchers have developed coated nanowires that bind to certain proteins that can indicate the presence of prostate cancer before conventional tests can. Other potential applications for nanowires include the early sensing of breast and ovarian malignancies. Nanowires are so small that doctors could one day implant them into the body as permanent health detectives that continuously monitor molecular levels.

Nanocantilever

The honeycomb mesh behind this tiny carbon cantilever is the surface of a fly's eye. Cantilevers are beams anchored at only one end. In the nanoworld, they function as sensors ideal for detecting the presence of extremely small molecules in biological fluids. Arrays of nanocantilevers coated with antibodies, for example, will bend from the changes in surface tension when substrates that signal a malignancy bind to it. Simply by monitoring whether or not such nanocantilevers are bent, specialists may someday be able to identify the presence of cancer molecules that today are difficult to detect.

Nanoshells

Nanoshells are hollow silica spheres covered with gold. Scientists can attach antibodies to their surfaces, enabling the shells to target certain cells such as cancer cells. In mouse tests, Naomi Halas's research team at Rice University directed infrared radiation through tissue and onto the shells, causing the gold to superheat and destroy tumor cells while leaving healthy ones intact.

Technicians can control the amount of heat with the thickness of the gold and the kind of laser. Nanoshells could one day also be filled with drug-containing polymers. Heating them would cause the polymers to release a controlled amount of the drug. Human trials using gold nanoshells are slated to begin in a couple of years.

Quantum Dots

Quantum dots are miniscule semiconductor particles that can serve as signposts of certain types of cells or molecules in the body. They can do this because they emit different wavelengths of radiation depending on the type of cadmium used in their cores: cadmium sulfide for ultraviolet to blue, cadmium selenide (seen here) for most of the visible spectrum, and cadmium telluride for the far red and near-infrared. (A dot's size determines its precise color within each range.) A polymer coating enables researchers to attach molecules such as antibodies that will seek out and attach to tumors and other targeted cells. The coating also shields nearby cells from the cadmium's toxicity. The different colors of quantum dots provide a powerful tool for labeling and monitoring multiple cells and molecules simultaneously.

CANCER TREATMENT WITHOUT HARMING HEALTHY TISSUE

Scientists at Stanford University have developed a new laser therapy that destroys cancer cells but leaves healthy ones unharmed. The new, non-invasive treatment is described in a study published in the Aug. 1 online edition of the Proceedings of the National Academy of Sciences (PNAS).

"One of the longstanding problems in medicine is how to cure cancer without harming normal body tissue," says Hongjie Dai, an associate professor of chemistry at Stanford and co-author of the study. "Standard chemotherapy destroys cancer cells and normal cells alike. That's why patients often lose their hair and suffer numerous other side effects. For us, the Holy Grail would be finding a way to selectively kill cancer cells and not damage healthy ones."

For the PNAS experiment, Dai and his colleagues used a basic tool of nanotechnology-carbon nanotubes, synthetic rods that are only half the width of a DNA molecule. Thousands of nanotubes could easily fit inside a typical cell.

"An interesting property of carbon nanotubes is that they absorb near-infrared light waves, which are slightly longer than visible rays of light and pass harmlessly through our cells," Dai says. But shine a beam of near-infrared light on a carbon nanotube, and the results are dramatic. Electrons in the nanotube become excited and begin releasing excess energy in the form of heat.

In the experiment, Stanford researchers found that if they placed a solution of carbon nanotubes under a near-infrared laser beam, the solution would heat up to about 158 degrees F (70 C) in two minutes. When nanotubes were placed inside cells and radiated by the laser beam, the cells were quickly destroyed by the heat. However, cells without nanotubes showed no effects when placed under near-infrared light.

"It's actually quite simple and amazing," Dai observes. "We're using an intrinsic property of nanotubes to develop a weapon that kills cancer."

Trojan Horse

To assure that only diseased cells were destroyed in the experiment, the scientists had to find a way to selectively deliver carbon nanotubes into cancer cells and not into healthy ones. Dai and his co-workers achieved this by performing a bit of biochemical trickery. Unlike normal cells, the surface of a cancer cell contains numerous receptors for a vitamin known as folate. The researchers decided to coat the nanotubes with folate molecules, which would only be attracted to diseased cells with folate receptors.

The experiment worked as predicted. Most of the folate-coated nanotubes ended up inside cancer cells, bypassing the normal cells-like Trojan horses crossing the enemy line. Once the nanotubes were planted inside, the researchers shined the near-infrared laser on the cancer cells, which soon heated up and died. "Folate is just an experimental model that we used," Dai says. "In reality, there are more interesting ways we can do this. For example, we can attach an antibody to a carbon nanotube to target a particular kind of cancer cell."

One example is lymphoma, or cancer of the lymphatic system. Like many cancers, lymphoma cells have well-defined surface receptors that recognize unique antibodies. When attached to a carbon nanotube, the antibody would play the role of a Trojan horse. Dai and Dean Felsher, a lymphoma researcher in the Stanford School of Medicine, have begun a collaboration using laboratory mice with lymphoma. The researchers want to determine if shining near-infrared light on the animal's skin will destroy lymphatic tumors, while leaving normal cells intact.

"It's a really interesting idea," says Felsher, an assistant professor of medicine and of pathology. "For a long time people have thought about ways to target cancer cells, and this is a very promising technique."

Researchers at Rice University recently conducted a similar experiment on mice with cancerous tumors. Instead of carbon nanotubes, the Rice team injected the tumors with gold-coated nanoshells and exposed the animals to near-infrared light for several minutes. The tumors disappeared within 10 days without damaging any healthy tissue.

Future Applications

Dai points out that the carbon nanotubes also can be delivered to diseased cells by direct injection. "In breast cancer, for example, there might come a time when we inject nanotubes into the tumor and expose the breast to near-infrared light," he says. This benign therapy could potentially eliminate months of debilitating chemotherapy and radiation treatment, he adds.

"The laser we used is a 3-centimeter beam that's held like a flashlight," he notes. "We can take the beam and put anywhere we want. We can shine it on a local area of the skin or inside an internal organ using a fiber-optic device."

Dai has applied for a patent on the procedure through Stanford's Office of Technology Licensing (OTL). He also has patented another technique that's designed to deliver molecules of DNA, RNA or protein directly into the cell nucleus to fight various infections and diseases. That method uses pulses of near-infrared light to shake the therapeutic molecule loose from the nanotube after they have entered the cell.

"Nanotechnology has long been known for its applications in electronics," Dai concludes. "But this experiment is a wonderful example of nanobiotechnology-using the unique properties of nanomaterials to advance biology and medicine."

Dai's graduate student, Nadine Wong Shi Kam, is lead author of the PNAS study. Other co-authors are Michael O'Connell, a former postdoctoral fellow in the Department of Chemistry, and graduate student Jeffrey A. Wisdom in the Department of Applied Physics.

The study was partly supported by the National Science Foundation Center on Polymer Interfaces and Macromolecular Assemblies, a research partnership among Stanford, IBM Almaden Research Center, University of California-Davis and University of California-Berkeley.

NANOTECHNOLOGY KILLS CANCER CELLS

Nanotechnology has been harnessed to kill cancer cells without harming healthy tissue. The technique works by inserting microscopic synthetic rods called carbon nanotubules into cancer cells. When the rods are exposed to near-infra red light from a laser they heat up, killing the cell, while cells without rods are left unscathed. Researcher Dr Hongjie Dai said: "One of the longstanding problems in medicine is how to cure cancer without harming normal body tissue." "Standard chemotherapy destroys cancer cells and normal cells alike. That's why patients often lose their hair and suffer numerous other side effects." "For us, the Holy Grail would be finding a way to selectively kill cancer cells and not damage healthy ones."

Many in Cell

The carbon nanotubules used by the Stanford team are only half the width of a DNA molecule, and thousands can easily fit inside a typical cell. Under normal circumstances near-infra red light passes through the body harmlessly. But the Stanford team found that if they placed a solution of carbon nanotubules under a near-infra red laser beam, the solution heated up to about 70°C in two minutes. They then placed the tubules inside cells, and found they were quickly destroyed by the heat generated by the laser beam.

Dr Dai said: "It's actually quite simple and amazing. We're using an intrinsic property of nanotubes to develop a weapon that kills cancer." The next step was to find a way to introduce the nantubules into cancer cells, but not healthy cells. The researchers did this by taking advantage of the fact that, unlike normal cells, the surface of cancer cells is covered with receptors for a vitamin known as folate. They coated the nanotubules with folate molecules, making it easy for them to pass into cancer cells, but unable to bind with their healthy cousins. Exposure to the laser duly killed off the diseased cells, but left the healthy ones untouched.

New Advanced Technique

The researchers believe it should be possible to refine the technique still further, for instance by attaching an antibody to a nanotubule to target a particular kind of cancer cell. Dr Emma Knight, of the charity Cancer Research UK, said: "Nanotechnology has a lot to offer biomedical science, and the results of this paper suggest yet another way in which it may help in the fight against cancer. "However, this work is still at a very early stage. The researchers have shown that near-infra red light can cause nanotubes to produce heat that can kill cancer cells. "But their work so far has focused on cells that have been grown in culture in the laboratory. "Further research will be crucial to see whether these effects can be reproduced in the more complex environment of a tumour and, ultimately, the human body."

Future of Nanotechnology in Cancer Treatment

Experiments on mice have shown promise for the future of nanotechnology in treating cancer. The research brings doctors one step closer to being able to inject patients with nanoparticles that bore inside tumours and release powerful doses of cancer-killing drugs while leaving the rest of the body unscathed. After seeing how the mice were cured of human prostate cancer with the technology, cancer specialists gathered at the European Cancer Conference in Paris praised the work as impressive and said they had high hopes for its application to patients.

"There are a lot of candidates for intelligent carriers, and these nanoparticles are among them," said Dr. Gordon McVie, a professor at the European Institute of Oncology in Milan, Italy, who was not involved with the research. Previous designs of nanoparticles have used antibodies to zone in on cancer cells.

"The body's immunodefence system can create antibodies to the therapeutic antibodies, deactivate them and prevent the antibody binding to the right cancer cells.

This looks like a step forward," Kerr said. Nanotechnology is the science of manipulating matter smaller than 100 nanometres and taking advantage of properties that are present only at that level, such as conductivity. A nanometre is one-billionth of a metre, or about one-millionth the size of a pin head. The prefix comes from "nanos," the Greek word for dwarf.

Nanotech has been around for several decades, but only now is its potential starting to be realized. Medicine is expected to be one of the fields to benefit most from the technology. In cancer, it is hoped the technology will allow for more precisely targeted drugs and surgery and less toxic chemotherapy.

The study, conducted by scientists at Harvard Medical School and the Massachusetts Institute of Technology, which are pioneering cancer nanotechnology, involved engineering nanoparticles embedded with the cancer drug Taxotere. The particles were then injected directly into human tumours created from prostate cancer cell lines and implanted into the flanks of mice. The mice were watched for 100 days.

The technology being tested involves a nanoparticle made of a hydrogen and carbon polymer with bits of drug bound up in its fabric and attached to a chemical that hones in on cancer cells. The polymer gradually dissolves, exposing the nuggets of drug little by little. The mice were divided into five groups, including one that had their tumours injected with ineffective saltwater. A second group died after injections of a nanoparticle containing no drugs.

NANOTECHNOLOGY FOR TARGETED CANCER THERAPY

Cancer nanotechnology is currently under intense development for applications in cancer imaging, molecular diagnosis and targeted therapy. The basic rationale is that nanometer-sized particles, such as biodegradable micelles, semiconductor quantum dots and iron oxide nanocrystals, have functional or structural properties that are not available from either molecular or macroscopic agents. When linked with biotargeting ligands, such as monoclonal antibodies, peptides or small molecules, these nanoparticles are used to target malignant tumors with high affinity and specificity. In the 'mesoscopic' size range of 5-100 nm in diameter, nanoparticles also have large surface areas and functional groups for conjugating to multiple diagnostic (e.g., optical, radioisotopic or magnetic) and therapeutic (e.g., anticancer) agents. Recent advances have led to multifunctional nanoparticle probes for molecular and cellular imaging, nanoparticle drugs for targeted therapy, and integrated nanodevices for early cancer detection and screening. These developments have opened exciting opportunities for personalized oncology in which cancer detection, diagnosis and therapy are tailored to each individual's molecular profile, and also for predictive oncology, in which genetic/molecular information is used to predict tumor development, progression and clinical outcome.

Cancer Therapy with No side-effects

Most of you know at least one person who has been affected by cancer and treated by chemotherapy, suffering from side-effects such as hair loss or nausea. This is because chemotherapy attacks both healthy and unhealthy cells in the whole body. Now, Australian researchers are using nanotechnology to offer chemotherapy relief. They've created 'nano-cells' from 'inert' bacteria-meaning they can't reproduce-which can deliver potent drugs exactly where they're needed. As this new therapy allows to target very precisely the tumors, the amounts of drugs are much smaller and the harmful side-effects of chemotherapy will be avoided in the future. This method could be used for a wide variety of cancers and human clinical trials should start by the end of this year. Here are some short excerpts of the Agençe France-Presse report.

Researchers there have used bacterial cells stripped of reproductive powers to develop receptacles capable of carrying any chemotherapy drug. These 'nano-cells', which are about one-fifth the size of of normal human cells, are then tagged with antibodies, which are attracted to cancerous tumours. Once the nano-cell hits the cancer and attaches, the drug is released directly into the malignant growth.

As said Jennifer MacDiarmid of Sydney-based biotechnology company EnGeneIC, "Your hair wouldn't fall out, you wouldn't throw up... some chemotherapy is life-threatening in itself." And she added that with their new method, "There is no other system where you can get so much drug concentrated into a little parcel."

In "Cancer therapy cuts side effects," BBC News adds more details about the future of this method.

It is hoped the technique could treat tumours in the breast, ovaries, colon and lungs. So far it has been used in mice and dogs, and the researchers plan to begin human trials shortly. In tests, the technique reduced tumours in animals without side effects-and by using only a very small amount of drug. Dr MacDiarmid said the treatment could potentially be used on any solid tumours. In "'Mini-cells' could stop side-effects of chemotherapy," New Scientist asks the question: "Could bacteria be the future of cancer therapy?" And it describes what the researchers have done with bacteria.

When bacteria divide, they normally do so at their centres. But Himanshu Brahmbhatt and Jennifer MacDiarmid, of EnGeneIC in Sydney, Australia, have found a way of forcing them to divide at their ends, producing small buds of cytoplasm each time. They have also discovered that a range of different drugs could be packaged into these particles. These "mini bacteria", or EnGeneIC Delivery Vehicles (EDVs) as the company has dubbed them, are cheap and easy to produce, and can be used as targeted drug delivery vehicles.

"They look like bacteria but they have no chromosomes and are non-living," says MacDiarmid. "And because they have a rigid membrane they don't break down when injected, so they carry their payload happily to the target site."

Systemic administration of chemotherapeutic agents results in indiscriminate drug distribution and severe toxicity. Here we report a technology potentially overcoming these shortcomings through encapsulation and cancer cell-specific targeting of chemotherapeutics in bacterially derived 400 nm minicells. We discovered that minicells can be packaged with therapeutically significant concentrations of chemotherapeutics of differing charge, hydrophobicity, and solubility.

These findings have also recently been presented by Himanshu Brahmbhatt, Joint CEO, EnGeneIC Pty Ltd, at the RNAi 2007 conference held in Boston, Massachusetts, on May 3, 2007. By scrolling down through this page, you'll discover the abstract of the presentation named "Turning siRNA into a cancer therapeutic: Bacterially-derived non-living nanoparticles for the targeted delivery of siRNA and shRNA for potent anti-tumor effects in vivo." [Note: siRNA is a shortcut for Small interfering RNA, as described by Wikipedia.] Below is a comment about how these nanoparticles, dubbed as EDVs (EnGeneIC Delivery Vehicles), are acting.

We discovered that EDVs can be packaged with therapeutically significant concentrations of siRNAs as well as chemotherapeutic drugs of differing charge, hydrophobicity and solubility. Targeting of EDVs via bispecific antibodies (BsAbs) to receptors on cancer cell membranes results in endocytosis, intracellular degradation and release of the payload. Intact and fully functional siRNAs are released intracellularly to achieve target mRNA knockdown resulting in highly significant anti-tumor effects as well as drug resistance reversal.

3-D Model of Breast Cancer in the Lab

The reaserchers of U.K. have built a 3-D version of breast cancer in a test tube. Their model contains cells from normal and cancerous breast tissue. The researchers used a collagen gel to form 3-D structures to create structures similar to the ones find in a woman breast. So far, they focused on a common pre-cancerous condition known as 'ductal carcinoma in situ' (DCIS). With this model, they hope to reduce experiments done on animals such as mice. In fact, these experiments are not always useful because similarities can be poor between mice and humans. Now it remains to be seen if this model will be endorsed by the scientific community.

Before going further, here is a definition of what is DCIS: "Ductal carcinoma in situ (DCIS) is a condition that can occur in several different forms. In all of these, there is a change in the cells lining the milk ducts of the breast. Under a microscope, a normal duct shows a lining of a single layer of cells, but in DCIS, many more cells can be seen. [...] These extra cells in the duct show some of the abnormalities of cancer cells, but they have not spread outside the duct (in situ means 'in place'). For this reason, DCIS is not considered to be 'true' or invasive cancer." Below is a diagram of a breast cross section showing ducts without and with DCIS (Credit for image and caption: The Cancer Council Victoria, Australia)

This project has been led by Louise Jones, professor of breast pathology at the Institute of Cancer at Queen Mary, University of London. Here are some of our explanations about this project.

Jones said "that they needed to develop a test tube model that was more complex than a layer of cells in a Petri dish. Instead the team grew three types of cells from normal and cancerous breast tissue. And using a collagen gel to form three-dimensional structures they have been able to create structures which closely resemble the ways the cells would be organised in the breast."

But how this 3-D Model Help in Cancer Research?

Professor Jones said working out why the cells behave like this could help doctors decide which patients with early cancer to treat and pave the way for developing new treatments. The model could also replace the need for some experiments which are done in animals such as mice, which can be implanted with human tumours. This is where my interest was growing. Is this research done for the interest of women or mice? In fact, this research is being funded by the Dr Hadwen Trust for an amount of £105,000. And "the Dr Hadwen Trust is the UK's leading medical research charity that funds and promotes exclusively non-animal research techniques to replace animal experiments." So let's look at the press release they issued on May 10, 2007, "Breast cancer in a test tube could replace animal tests."

Breast cancer research is often conducted in mice, with up to 400 mice used in a series of tests. Each animal is likely to endure up to eight separate tumour implants followed by painful drug injections into the abdomen, and distressing, serial 'harvesting' of subsequent tumours. As well as causing suffering, these mouse experiments are potentially misleading, meaning a human-based research approach is vital to improve accuracy.

As says Nicky Gordon, Science Officer for the Dr Hadwen Trust: "As a medical research charity we aim to alleviate human suffering, but we believe that medical excellence can and should be achieved without causing animals suffering too. With breast cancer, there is an urgent need to move away from animal research models because their similarity to human cancer can be so poor.

ADVANCE LIGHT-ACTIVATED CANCER THERAPY

Photodynamic therapy (PDT) is a powerful approach for destroying tumors, one that uses light and a light-sensitive chemical to rapidly kill cells. Unfortunately, PDT is associated with significant side effects that largely result from the fact that light-sensitive chemicals, or photosensitizers, distribute themselves throughout the body and thus can damage healthy as well as malignant cells. Another limitation arises because the frequency of light needed to excite the photosensitizers can only pass through a few millimeters of skin.

To reduce or even eliminate PDT's therapy-limiting side effects, researchers are attempting to increase uptake of photosensitizers by tumors and decrease uptake by healthy cells, and to do that investigators are relying heavily on nanoparticles. In a new report, investigators at the State University of New York in Buffalo describe a novel nanoparticle that encapsulates a photosensitizer and a special dye molecule that can absorb light capable of traveling deep into the body and transfer that light energy to the photosensitizer. Tests with cultured cancer cells show that this nanoparticle kills tumor cells.

Paras Prasad, Ph.D., principal investigator of the Multifunctional Nanoparticles in Diagnosis and Therapy of Pancreatic Cancer program, one of National Cancer Institute-funded Cancer Nanotechnology Platform Partnerships, led this research effort. The results of this work appear in the

Journal of the American Chemical Society. The main component of this new system is an organically modified silica nanoparticle that Prasad's group has been developing for several years. Within this nanoparticle, the investigators encapsulated a novel dye molecule that can absorb and emit light in spite of being packed closely to other dye molecules; the optical performance of most dye molecules drops precipitously when they are in close contact with other dye molecules. The nanoparticles also contain a photosensitizer that efficiently absorbs light emitted by this particular dye. This photosensitizer, known as HPPH, recently began to be tested in human clinical trials.

Initial tests with this nanoparticle formulation showed that it generated cell-killing reactive oxygen when irradiated with near-infrared light, the portion of the light spectrum that passes through skin and other tissues easily. Additional experiments showed that tumor cells took in these nanoparticles over a 3-hour period and that when the cells were subsequently irradiated with near-infrared light, the cells died from the internally generated reactive oxygen.

In a second study, published in the journal Molecular Pharmaceutics, the investigators showed that nanocrystals of HPPH could also trigger reactive oxygen production in tumor cells. The purpose of this study was to determine if it was possible to create a formulation of HPPH that is stable in water yet retains its cell-killing activity. HPPH belongs to a category of drugs, which includes many anticancer agents, that are poorly soluble in water. Pure drug nanocrystals dispersed readily in water and were taken up by tumor cells.

Bibliography

Andrew, N.: *Foundations of Nanomechanics: from Solid-state Theory to Device Applications,* Berlin, New York, 2003.

Attebery, B.: *Nanoculture: Implications of the New Technoscience,* Bristol, Intellect Books, 2004.

Berube, D.M.: *The Rhetoric of Nanotechnology,* Amsterdam: IOS Press, 2004.

Carroll, J.S.: *Social Science Research Methods for Assessing Societal Implications of Nanotechnology,* Dordrecht, Kluwer, 2001.

Charles, P., and Frank J. Owens: *Introduction to Nanotechnology,* Hoboken, J. Wiley, 2003.

Daniel Ratner: *Nanotechnology: a Gentle Introduction to the Next Big Idea,* Upper Saddle River, Prentice Hall, 2003.

Drexler, K. Eric: *Engines of Creation: The Coming Era of Nanotechnology,* New York, Anchor Books, 1986.

Etzkowitz, H.: *Nano-Science and Society: Finding a Social Basis for Science Policy,* Dordrecht, Kluwer, 2001.

Fleeter, Rick: *The Logic of Microspace,* El Segundo, CA: Microcosm Press, 2000.

Frank J. Owens: *Introduction to Nanotechnology,* Hoboken, J. Wiley, 2003.

Fumento, Michael: *Bioevolution: How Biotechnology is Changing Our World,* San Francisco, California: Encounter Books, 2003.

Gaskell, G: *Biotechnology - the Making of a Global Controversy,* Cambridge, Cambridge University Press, 2002.

George S. Paul: *Beyond Humanity: Cyber Evolution and Future Minds,* Roackland, Charles River Media, 1996.

Glimell, H.: *Dynamics of the Emerging Field of Nanoscience,* Dordrecht, Kluwer, 2001.

Gregory, L.: *Nanotechnology,* New York, Springer, 1999.

Hayles, N.K. : *Nanoculture: Implications of the New Technoscience,* Bristol, Intellect Books, 2004.

Hessenbruch, A.: *Nanotechnology and the Negotiation of Novelty,* Amsterdam, IOS Press, 2004.

James Lewis: *Prospects in Nanotechnology: Toward Molecular Manufacturing,* New York, John Wiley & Sons, 1995.

Jin Z. Zhang: *Self-assembled Nanostructures,* New York, Kluwer Academic/ Plenum, 2003.

Kalish, R., and J. E. Butler: *Optical Characterization of Boron-doped Nano-crystalline Diamond,* Washington, Naval Research Laboratory, 2002.

Katz, E., R. Lovel, W. Mee and F. Solomon: *Citizens' Panel on Nanotechnology: Report to Participants,* Melbourne, CSIRO Minerals, 2005.

Kurzweil, Ray: *The Age of Spiritual Machines*, New York, Penguin Books, 1999.

Mark, A., and Daniel Ratner: *Nanotechnology: a Gentle Introduction to the Next Big Idea,* Upper Saddle River, Prentice Hall, 2003.

Meyer, M.: *Socio-economic Research on Nanoscale Science and Technology: A European Overview and Illustration,* Dordrecht, Kluwer, 2001.

Moravec, Hans: *Mind Children: The Future of Robot and Human Intelligence*, Cambridge, Harvard University Press, 1988.

Nigel, M. De S. Cameron: *Human Dignity in the Biotech Century: A Christian Vision for Public Policy*, Downers Grove, InterVarsity Press, 2004.

Norio Taniguchi: *Nanotechnology: Integrated Processing Systems for Ultra-precision and Ultra-fine Products,* New York, Oxford University Press, 1996.

Orth, R. N.: *Nanometer Scale Antibody Patterning for Directed cell Immobilization and Stimulation,* Ithaca, Cornell University, 2002.

Parens, Erik: *Enhancing Human Traits: Ethical and Social Implications,* Washington, D.C., Georgetown University Press, 1998.

Postman, Neil: *Technopoly: The Surrender of Culture to Technology*, New York, Vintage Books, 1992.

Pressman, J.: *Nano Narrative: A Parable from Electronic Literature,* Bristol, Intellect Books, 2004.

Richard, Hewlett: *A Policy Framework for Developing a National Nanotechnology Program*, Blacksburg, University Libraries, 1998.

Rosalyn, W. : *Encyclopedia of Science, Technology and Ethics*, Detroit, Thomson-Gale, 2005.

Sarewitz, D.: *Nanotechnology and Societal Transformation*, Dordrecht, Kluwer, 2001.

Schummer, J.: *Interdisciplinary Issues in Nanoscale Research*, Amsterdam: IOS Press, 2004.

Seeman, N.: *DNA Nanotechnology for Massive Information Storage,* New York, New York University, 2001.

Smith, H. A.: *Nanotechnology Infrared Optics for Astronomy Missions,* Washington, National Aeronautics and Space Administration, 2002.

Srivastava, D.: *Nanotechnology based Materials and Devices for Health Care*, Moffett Field, Computer Sciences Corp., 2002.

Stephen M. Goodnick: *Transport in Nanostructures,* Cambridge, Cambridge University Press, 1997.

Suchman, M.: *Envisioning Life on the Nano-Frontier*, Dordrecht, Kluwer, 2001.

Tenner, E.: *Nanotechnology and Unintended Consequences*, , Dordrecht, Kluwer, 2001.

Tomellini, R.: *Nanotechnology: Revolutionary Opportunities and Societal Implications*, *Workshop, Lecce*, Luxemburg: Office for Official Publications of the European Communities, 2002.

Weil, V.: *Ethical Issues in Nanotechnology*, Dordrecht, Kluwer, 2001.

Index

N

P

S

T

❑❑❑